Erika Hauk
Investigation of a
Predictive Loss Balancing Method for the
Three-Level Active Neutral Point
Clamped Voltage Source Converter

I0131715

TUD*press*

Erika Hauk

Investigation of a Predictive Loss Balancing Method for the Three-Level Active Neutral Point Clamped Voltage Source Converter

TUDpress
2016

Die vorliegende Arbeit wurde am 10. Dezember 2014 an der Fakultät Elektrotechnik und Informationstechnik der Technischen Universität Dresden als Dissertation
eingereicht und am 09. Juli 2015 verteidigt.

Vorsitzender:
Univ.-Prof. Dr.-Ing. Wilfried Hofmann, Technische Universität Dresden

Gutachter:
Prof. Dr.-Ing. Steffen Bernet, Technische Universität Dresden
Prof. Dr.-Ing. Andreas Lindemann, Otto-von-Guericke-Universität Magdeburg

Bibliografische Information der Deutschen Nationalbibliothek
Die Deutsche Nationalbibliothek verzeichnet diese Publikation in der
Deutschen Nationalbibliografie; detaillierte bibliografische Daten sind
im Internet über http://dnb.d-nb.de abrufbar.

Bibliographic information published by the Deutsche Nationalbibliothek
The Deutsche Nationalbibliothek lists this publication in the Deutsche
Nationalbibliografie; detailed bibliographic data are available in the
Internet at http://dnb.d-nb.de.

ISBN 978-3-95908-059-0

© 2016 w.e.b.
Universitätsverlag & Buchhandel
Eckhard Richter & Co. OHG
Bergstr. 70 | D-01069 Dresden
Tel.: 0351/47 96 97 20 | Fax: 0351/47 96 08 19
http://www.tudpress.de

Technische Universität Dresden

Investigation of a Predictive Loss Balancing Method for the Three-Level Active Neutral Point Clamped Voltage Source Converter

Erika Hauk

von der Fakultät Elektrotechnik und Informationstechnik der
Technischen Universität Dresden

zur Erlangung des akademischen Grades eines

Doktoringenieurs

(Dr.-Ing.)

genehmigte Dissertation

Vorsitzender:	Prof. Dr.-Ing. Hofmann	Tag der Einreichung: 10.12.2014
Gutachter:	Prof. Dr.-Ing. Bernet	Tag der Verteidigung: 09.07.2015
	Prof. Dr.-Ing. Lindemann	

Acknowledgements

I would like to express my sincere gratitude to my advisor Prof. Bernet, for his scientific advice and knowledge he has provided throughout my doctorate years. Besides my advisor, I also have to thank the members of my PhD committee, Prof. Lindemann, Prof. Janschek, and Prof. Hofmann, for their helpful advice and suggestions in general. I also thank Dr. Weber for his help and teaching.

I would further like to thank all my colleagues and friends from Vietnam, Germany, and Latin America for the technical discussion, for the friendship that I needed, and for the cultural impressions.

I must express my gratitude to my parents, Georgeta and Niculai Hauk, and my friend, Arturo Arias for their faith in me and for their continued support and encouragement.

Finally, I would like to thank the DAAD, not only for providing the funding this research, but also for giving me the opportunity to meet so many interesting people from all around the world.

CONTENTS

Variable definition

Variable	Meaning
a, b, c	phase node
$\underline{A}$, $\underline{B}$, $\underline{C}$	system's matrixes
$A_{sw}, B_{sw}, C_{sw}, B_{cond}$	fitting constants
C_{th}	thermal capacitance
d	duty cycle
d_D	duty cycle of the diode
d_T	duty cycle of the IGBT
E_{off}	turn-off energy of the IGBT
E_{on}	turn-on energy of the IGBT
E_{rec}	recovery energy of the diode
f_1	fundamental output frequency
f_{comm}	frequency of the common-mode dc offset
f_{sw}	switching frequency
g	cost function
h	harmonic order
I_C	dc collector current
$I_{C,n}$	nominal collector current of the IGBT
$I_{F,n}$	nominal forward current of the diode
i_M	neutral current
$i_{ph,rms}$	effective value of phase current
I_{TGQM}	maximum controllable turn-off current
k	discrete time point

m_a	modulation index
m_f	frequency modulation ratio
N	number of switching angles per quarter-fundamental period
n	number of dc-link voltage levels
p	the number of phases
P_{bl}	blocking losses
P_{cond}	conduction losses
P_{drive}	driving losses
$P_{loss,D}$	power losses of the diode
$P_{loss,T}$	power losses of the IGBT
P_{off}	turn-off losses
P_{on}	turn-on losses
P_{sw}	switching losses
P_{tot}	total power loss of a semiconductor
$P_{avg\,ALB}$, $P_{max\,PALB}$	decrease of the average junction temperature
$P_{max\,ALB}$, $P_{max\,PALB}$	decrease of the maximal junction temperature
$P_{\Delta\vartheta\,ALB}$, $P_{\Delta\vartheta\,PALB}$	decrease of the junction temperature ripple
R_{th}	thermal resistance
$R_{th,ca}$	thermal resistance case-to-ambient
$R_{th,jc}$	thermal resistance junction-to-case
$r_{\Gamma 0}$	slope resistance
t	time
t_1	on-time of the IGBT
$\underline{T}$	temperature differences as state variables
$t_{prediction}$	prediction interval
T_{sample}	sample period
T_{sw}	period of the switching frequency
S_{sn}	nominal switch power
S_w	power semiconductor
$u_{2L\text{-}SVM}$	reference waveform for 2L-SVM
$u_{2L\text{-}SVM}$	reference waveform for 3L-SVM
u_{aM}, u_{bM}, u_{cM}, u_{xM}	phase-midpoint output voltages, $x=a, b, c$
$\hat{u}_{xM,h}$	peak voltage of the harmonic component
u_{an}, u_{bn}, u_{cn}	phase-to-neutral point voltages
u_{ab}, u_{bc}, u_{ca}	line-to-line voltages

$U_{CE,n}$	on-state saturation voltage of the IGBT
$\hat{U}_{con,1}$	amplitude of the fundamental component of the control signals
U_{dc}	dc-link voltage
$U_{F,n}$	on-state voltage of the diode
u_{max}, u_{min}	maximum and minimum reference voltages
u_{nM}	common mode voltage
u_{off}	offset voltage
$u_{ref,a}, u_{ref,b}, u_{ref,c}$	three-phase sinusoidal reference signals
$u_{tri,up}, u_{tri,low}$	upper and lower carrier signal
u_x	Normalized reference voltage
$\hat{U}_{tri}$	amplitude of the triangular signal
U_{base}	voltage at which the losses were measured
U_{CE}	collector-emitter voltage
U_{comm}	commutation voltage
$U_{dc,n}$	nominal dc-link voltage
U_{DRM}	repetitive peak off-state voltage
$\vec{V}_{ref}$	reference phasor
$v_{\Gamma 0}$	the threshold voltage
x	zero state
x_{op0}	optimal zero state
Z_{th}	thermal impedance
α_k	switching angles
$\Delta \vartheta_j$	temperature ripple
ϑ_a	ambient temperature
$\vartheta_{j,avg}$	average temperature
$\vartheta_{j,max}$	maximal operating junction temperature
τ_{th}	thermal time constant

List of acronyms and names

Acronym/ Name	Meaning
ALB	active loss balancing
ANPC	active neutral point converters
GTO	gate turn-off thyristors
HB	H-bridge
IGCT	integrated gate commutated thyristors
IGBT	insulated gate bipolar transistors
PP	press-pack
FIT	failure in time, i.e. , in 109 hours of operation
FLC	flying capacitor
PALB	predictive active loss balancing
PD	phase disposition
PEBB	power electronics building blocks
PWM	pulse width modulation
ML	multi level
MV	medium voltage
NPC	neutral point converters
NTV	nearest three space vectors
SCHB	cascaded H-bridge converter
SHE	selective harmonic elimination
SVM	space vector modulation
VSC	voltage source converter
T_{out}	outer switches T_{11}, T_{21}, T_{31} and T_{14}, T_{24}, T_{34}
D_{out}	outer diodes D_{11}, D_{21}, D_{31} and D_{14}, D_{24}, D_{34}
T_{in}	inner switches T_{12}, T_{22}, T_{32} and T_{13}, T_{23}, T_{33}
D_{in}	inner diodes D_{12}, D_{22}, D_{32} and D_{13}, D_{23}, D_{33}
T_{NPC}	NPC switches T_{15}, T_{16}, T_{25}, T_{26}, T_{35} and T_{36}
D_{NPC}	NPC diodes D_{15}, D_{16}, D_{25}, D_{26}, D_{35} and D_{36}

Abstract

Nowadays, the application of Medium Voltage (MV) drives has grown due to the increasing demand of high efficiency and energy saving in industry, traction, power generation, power transmission, and distribution. Among the MV multilevel converters available on the market, the Three-Level Neutral Point Clamped Voltage Source Converter (3L-NPC VSC) has emerged as one of the most important topologies. One important disadvantage of this topology is the unequal power loss distribution and consequently, an unsymmetrical junction temperature distribution. In order to ensure that the most stressed power switches do not exceed the maximal allowable junction temperature, the converter output power and switching frequency have to be limited. To overcome this drawback, the Three-Level Active Neutral Point Clamped Voltage Source Converter (3L-ANPC VSC) was developed. The 3L-ANPC VSC provides additional switching states and commutations, which enable an advantageous distribution of the semiconductor losses. The loss distribution can be adjusted by the selection of an optimal zero state using a loss balancing method.

There are two different loss balancing methods evaluated in this work: the already known Active Loss Balancing (ALB) method and a new method, the predictive active loss balancing (PALB) method. The efficiency of the balancing methods is evaluated concerning the junction temperature and the maximum achievable phase current. The 3L-ANPC VSC is investigated at grid frequency, at low fundamental frequency, and at zero speed. Moreover, the thermal behavior of the 3L-ANPC VSC is analyzed for two IGBT packaging types: module and press-pack (PP), due to the different thermal coupling and thermal restrictions.

The use of the balancing methods applied to the 3L-ANPC VSC present outstanding results in all critical operation points as compared to the performance of the 3L-NPC VSC. Thus, the phase current and the output power range can be increased without exceeding the thermal limitation of the semiconductors. Furthermore, the loss balancing methods reduce the junction temperature ripple and consequently, the thermo-dynamical stress of the semiconductor. Therefore, the lifetime expectancy of the semiconductors and the reliability of the converter are increased. The most important improvements were obtained by the PALB method applied to the 3.3 kV 3L-ANPC VSC using 4.5 kV press-pack IGBTs at small modulation index, small output frequency, and zero speed in comparison to the ALB method. The reason is that the PALB method explicitly considers the conduction losses that are especially important at small modulation depth.

Kurzfassung

Aufgrund der Notwendigkeit von Energieeinsparungen und der Notwendigkeit einer verbesserten Wandlung und Dosierung von Energie in der Industrie, Traktion, Energieerzeugung, Energieübertragung und - verteilung steigt die Bedeutung von Mittelspannungsantrieben kontinuierlich. Eine heute auf dem Markt für Mittelspannungsstromrichter weit verbreitete Mehrpunkt-Stromrichtertopologie ist der Dreipunkt-Stromrichter (*Three-Level Neutral Point Clamped Voltage Source Converter*, 3L-NPC VSC). Ein wesentlicher Nachteil dieser Schaltung ist die ungleiche Verlustverteilung in den Leistungshalbleitern und infolgedessen eine unsymmetrische Verteilung der Sperrschichttemperaturen. Um sicherzustellen, dass die am stärksten belasteten Leistungshalbleiter nicht ihre maximal zulässige Sperrschichttemperatur überschreiten, müssen Ausgangsleistung und Schaltfrequenz des Stromrichters begrenzt werden. Um diesen Nachteil zu überwinden, wurde der Aktive Dreipunkt-Stromrichter (*Three-Level Active Neutral Point Clamped Voltage Source Converter*, 3L-ANPC VSC) entwickelt. Dieser bietet zusätzliche Schaltzustände und Kommutierungspfade, welche eine Symmetrierung der Halbleiterverluste ermöglichen. Die Verlustverteilung kann durch die Auswahl eines geeigneten Nullzustands mittels eines Verlustsymmetrierungsverfahrens beeinflusst werden.

Im Rahmen dieser Arbeit wurden zwei unterschiedliche Verlustsymmetrierungsverfahren untersucht: das schon bekannte „aktive Verlustsymmetrierungsverfahren" (*Active Loss Balancing method*, ALB) und das neue „prädiktive aktive Verlustsymmetrierungsverfahren" (*Predictive Active Loss Balancing method* (PALB)). Die Effizienz der Verlustsymmetrierungsverfahren wird auf Basis der Kriterien Sperrschichttemperatur und maximal erreichbarer Phasenstrom bewertet. Die Auswertung erfolgt für zwei Beispielstromrichter bei Netzfrequenz, bei niedriger Ausgangsfrequenz und bei einer Ausgangsfrequenz von Null. Ferner werden wegen unterschiedlicher Wärmekopplungen und diverser thermischer Einschränkungen zwei verschiedene IGBT-Gehäusetypen – Modul und Press-Pack – betrachtet.

Der 3L-ANPC VSC bietet bei Verwendung beider Verlustsymmetrierungsverfahren hervorragende Ergebnisse in allen kritischen Betriebspunkten im Vergleich zum 3L-NPC VSC. Der Phasenstrom und infolgedessen der Ausgangsleistungsbereich können erhöht werden, ohne dass es zu einer Überschreitung der thermischen Grenzen der Leistungshalbleiter kommt. Ferner reduzieren die Verlustsymmetrierungsverfahren die Welligkeit der Sperrschichttemperaturen und damit die thermomechanischen Belastungen der Leistungshalbleiter. Daraus folgt eine Erhöhung der Lebensdauer der Halbleiter und der Zuverlässigkeit des Stromrichters. Das PALB-Verfahren zeigt die wichtigsten Verbesserungen gegenüber dem ALB-Verfahren bei kleinem Modulationsindex und geringen Ausgangsfrequenzen. Der Grund dafür ist, dass das PALB-Verfahren explizit die Durchlassverluste berücksichtigt, die bei kleinem Modulationsindex wesentlich sind.

1 Introduction

Nowadays, the importance of Medium Voltage (MV) drives is growing due to the increasing demand of a dynamic and high efficient energy conversion in industry, traction, power generation, power transmission, and distribution [1], [2]. The development of high voltage semiconductors with turn-off capability, improved converter designs and control methods, and new topologies have enabled a substantial development of MV multilevel converters in the last decades.

Among the MV multilevel converters available on the market, the Three-Level Neutral Point Clamped Voltage Source Converter (3L-NPC VSC) has emerged as one of the most important topologies [3]. The 3L-NPC VSC presents advantages like extended output voltage range and lower harmonic content compared to the Two-Level Voltage Source Converter. Furthermore, it is characterized by a low part count, modularity, robustness, high reliability, and availability which have made this topology competitive at medium and high- power applications [4], [5].

The 3L-NPC VSC can be mainly found featuring Insulated Gate Bipolar Transistors (IGBTs) or Integrate Gate Commutated Thyristors (IGCTs). The 3L-NPC VSC using IGBT modules is widely used for pumps, fans, extruders, marine drives, and mill drives with power ratings from 0.8 MVA to 10 MVA. For an increased output power up to 32 MVA, configurations using IGCTs or press-pack IGBTs are employed [5]. These converters are applied for hot and cold rolling mill drives, marine drives, compressors, pumps, the Transrapid, and interties [4].

The 3L-NPC VSC has also found applications in the emerging wind energy market [6], [7]. Furthermore, the 3L-NPC VSC has been adapted to a static compensator (STATCOM) for flexible AC transmission systems (FACTS), for reactive power compensation, and/- or voltage control of power systems [8].

1

1. Introduction

One main drawback of the topology is the unequal semiconductor loss distribution and consequently, the unsymmetrical junction temperature distribution [9]. The most stressed power semiconductor within the converter limits the switching frequency and the maximal output power. To overcome this issue, the 3L-ANPC VSC was introduced [10]. A junction temperature balance is achieved applying a loss balancing method. One successfully implemented method is the Active Loss Balancing (ALB) algorithm, which is reported in [9], [11]. The aim of the ALB method is to advantageously distribute the switching losses between different power semiconductors and thus, to achieve a better junction temperature balance. The highest junction temperature of the power semiconductors of one phase leg is substantially reduced, enabling an increase of the converter output power and/-or switching frequency [9].

The 3L-ANPC VSC is especially advantageous for applications where high torque at zero speed is required, for example in rolling mills. Furthermore, the 3L-ANPC converter represents an attractive option for high power applications, where the required output power can be achieved avoiding a serial/parallel connection of power devices or a parallel connection of two converters. Thus, key requirements like reduced size, weight, and costs per converter power can be achieved.

The aim of this work is to develop a new junction temperature balancing method for the 3L-ANPC VSC. The new balancing method is named Predictive Active Loss Balancing (PALB). The PALB method predicts the thermal behavior of the converter in order to advantageously distribute the switching and conduction losses by reducing the highest junction temperature of the power semiconductors of one phase leg. In order to evaluate the efficiency of the proposed balancing method, the PALB method is analyzed in comparison with the ALB method.

The thesis is structured in six chapters. Chapter 2 describes the important features of IGBTs and IGCTs. Furthermore, an overview of the commercially available MV drives is presented, including general features, advantages and disadvantages, range of operation, and a market

2

overview. Also, this chapter emphasizes the development of the 3L-NPC VSC and the 3L-ANPC VSC on the market of MV drives.

The function of the 3L-NPC VSC and the 3L-ANPC VSC are presented in Chapter 3. Their structures, switch states, commutations, and the applied pulse width modulation (PWM) schemes are described. The principles of the ALB method applied to the 3L-ANPC VSC are explained and evaluated in comparison to the conventional NPC converter regarding the loss and junction temperature distribution.

Chapter 4 presents a general method to calculate the conduction and switching losses of a power semiconductor. The Foster and Cauer thermal models of power semiconductors and the heat sink are presented. Furthermore, a general method to calculate the junction temperature using the Foster thermal model is described. Finally, the thermal behavior of the 3L-NPC VSC is analyzed at different operation points, including low fundamental frequency and zero speed.

Chapter 5 presents the description of the PALB method. The structure and the principle of operation of the PALB method are described. The performance of 3L-ANPC VSC applying the PALB and the ALB method is evaluated and compared to the conventional NPC converter. Furthermore, the balancing methods are compared regarding the maximal and average junction temperature, the distribution of switching and conduction losses, as well as the selection of the different zero switch states. The balancing methods are investigated at grid frequency, at low fundamental frequency, and at zero speed. Moreover, the efficiency of the balancing methods is evaluated in terms of maximal achievable output power and switching frequency.

The conclusions of this thesis are presented in Chapter 6.

3

2 State of the Art of Medium Voltage Converters

This chapter presents a review of the most commonly Medium Voltage (MV) semiconductors and converters used in the industry today. The review includes general features, advantages and disadvantages, range of operation and a market overview.

This chapter also emphasizes the applicability and development of the Three-Level Neutral Point Clamped Voltage Source Converter (3L-NPC VSC), followed by the Three-Level Active Neutral Point Clamped Voltage Source Converter (3L-ANPC VSC). Finally, the motivation of a new improved balancing method to overcome the unequal loss distribution in the 3L-NPC VSC is presented.

2.1 State of the Art of Medium Voltage Semiconductors

In recent years, the market of MV drive systems has been significantly expanded. Nowadays, numerous converter topologies and power semiconductor devices are available. The market of MV semiconductors with turn-off capability is dominated by Insulated Gate Bipolar Transistors (IGBT) and Integrated Gate Commutated Thyristors (IGCT). Today, these two power devices replace Gate Turn-off Thyristors (GTO) [2], [12]. The main reasons are the required bulky and expensive snubber circuit and the complex gate unit of GTOs.

The IGCT [13], [14] is a hybrid of an improved GTO structure and an extremely low inductive gate drive. IGCTs are available with or without reverse blocking capability. IGCTs with reverse blocking voltage capability are known as symmetrical IGCTs and are usually used in current source inverters. IGCTs without reverse blocking voltage capability are known as asymmetrical IGCTs and are used in applications where high reverse voltages do not occur. Asymmetrical IGCTs can be integrated with a reverse conducting diode in the same package (reverse conducting IGCTs) and are generally used in voltage source inverters [2].

4

The IGBT [13], [15] combines the advantages of the MOSFET (high impedance gate and simple gate-drive characteristics) with the low on-state voltage of the bipolar junction transistor (BJT) [12]. There are two main types of IGBT-chip structures: Punch-Through IGBTs (PT IGBTs) and Non-Punch-Through IGBTs (NPT IGBTs). Both chip structures present benefits and drawbacks and the choice of the IGBT type depends on the specific application [16]. Further developments of IGBTs are: Soft-Punch-Through IGBTs (SPT IGBTs), Trench-Gate IGBTs (Trench-IGBT), Injection-Enhanced IGBTs (IEGTs), and Carrier-Stored-Trench-Gate IGBTs (CSTBT-IGBTs) [15].There are two packaging types used for the MV semiconductors: the power module and the press-pack. GTOs and IGCTs are available only as press-pack devices, while IGBTs are manufactured in both packaging types. Table 2.1 presents the advantages and disadvantages of the module and the press-pack package. A qualitative comparison between IGBTs and IGCTs is briefly summarized in Table 2.2.

Table 2.1 Characteristics of module and press-pack [1], [13]

Module	Press-pack
Advantages	
isolated base plate that implies simple mounting and cooling by isolated heat sinks, low packaging costs, low expense for mounting	explosion-free in case of a short circuit (mostly), possibility of a redundant converter design, double-side cooling, higher reliability according to thermal and power cycling than the module;
Disadvantages	
reduced power cycling capability than the press-pack, undefined failure mode after short circuits (open or shorted terminals),possible explosion during a failure.	more expensive mounting and cooling than the module for applications with power lower than 3 MVA.

Press-pack IGBTs are applied mainly in self commutated High Voltage DC converters, where a redundant converter design is a main requirement. However, the press-pack IGBTs are considerably more

5

2. State of the Art of Medium Voltage Converter

complex structured and expensive to manufacture than press-pack IGCTs or module IGBTs [1].

Table 2.2 Characteristics of IGBT and IGCT [1], [2]

Item	IGBT	IGCT
Snubber circuits	Not required	Requirement due to limitation of di/dt during turn-on transient and limitation of short circuit peak current
Turn-on losses	High	Low (turn-on snubber)
Turn-off losses	Low	High
On-state losses	High	Low
Switching behavior	Adjustment of switching speed during turn-off and turn-on transient by the gate unit	Determined by the device structure, doping and clamp circuit
Short circuit current	Limited by operation in the active region; active turn-off capability of short circuit currents	Limited by the clamp circuit; safe discharge of the dc-link capacitor in the case of short circuit
Overvoltage limitation	By operation in the active region	By clamp circuit
Gate unit	Low power, simple, compact	Medium power, complex, integrated
Series connection	Simple by gate unit dv/dt control and/or active clamping	Complex (requirement of external voltage balancing networks)
Parallel connection	Simple (low parameter deviation, adjustment of the switching behavior by the gate unit)	Complex (requirement of external current balancing networks)
Behavior after destruction	Open circuit (mostly)	Short circuit
Reliability	About 100 FIT per device (based on field failure rates)	About 100 FIT per device (based on field failure rates)

Table 2.3 presents a market overview of the available power semiconductors with turn-off capability. The nominal switch power S_{sn} is a measure for the maximum device power handling capability [1]. In the case of IGCTs, the nominal switch power is calculated as $S_{sn}=I_{TGQM} \cdot V_{DRM}$, where I_{TGQM} is the maximum controllable turn-off current and V_{DRM} is the

6

repetitive peak off-state voltage. In the case of IGBTs, the nominal switch power is calculated by $S_{sn} = V_{CE} \cdot I_C$, where I_C is the nominal collector current and V_{CE} is the collector-emitter voltage. Although IGBTs are typically able to turn-off twice the rated current ($2 \cdot I_C$) the power ratings of the majority of IGBTs is still lower compared to IGCTs.

Table 2.3 Device rating and package types of available MV semiconductor devices

Power Devices	Manufacturer	Voltage Ratings (V_{DRM}/V_{CE})	Current Ratings (I_{TGQM}/I_C)	Switch Power S_{Sn} in MVA	Case
IGBT	INFINEON	3300 V	400–1500 A	4.95	module
		6500 V	200–750 A	4.875	module
	MITSUBISHI	3300 V	400–1500 A	4.95	module
		4500 V	400–900 A	4.05	module
		6500 V	200–600 A	3.9	module
	HITACHI	3300 V	800–1500 A	4.95	module
		4500 V	600–1200 A	5.4	module
		6500 V	400–750 A	4.875	module
	TOSHIBA	3300 V	400–1200 A	3.96	PP
		4500 V	1200–2100 A	9.45	PP
	ABB	3300 V	800–1500 A	4.95	PP
		6500 V	400–750 A	4.875	PP
	POWEREX	3300 V	800–1500 A	4.95	PP
		4500 V	400–900 A	4.05	PP
		6500 V	200–600 A	3.9	PP
	WESTCODE	4500 V	160–2400A	10.8	PP
IGCT	ABB	4500 V	4000–5000 A*	24.75	PP
		4500 V	630–2200 A**	9.9	PP
		5500 V	520–1800 A**	9.9	PP
		6500 V	3800 A*	24.7	PP

*: Asymmetric blocking device **: Reverse conducting device

Figure 2.1 synthesizes the maximum nominal voltage and maximum turn off current ratings of the semiconductors described above. Infineon, Hitachi, ABB, and Powerex produce IGBT modules with blocking voltages up to 6.5 kV and with current ratings up to 1.5 kA.

7

2. State of the Art of Medium Voltage Converter

Toshiba and Westcode offer press-pack IGBTs with blocking voltages up to 4.5 kV and currents up to 2.4 kA ($I_{C,n}$). ABB produces asymmetric and symmetric press-pack IGCTs with maximum blocking voltages of 6.5 kV and currents of 5.5 kA. Noteworthy is the ABB 10 kV asymmetric IGCT [17], which would enable a converter voltage of 6–7.2 kV, without using a series connection of power devices in a 3L-NPC VSC.

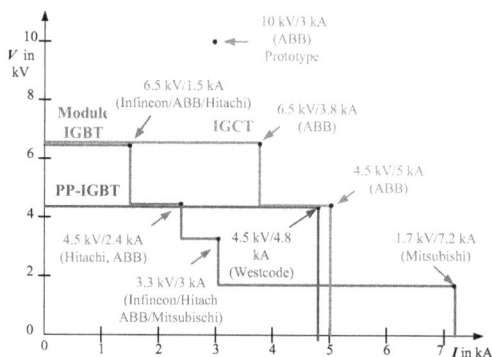

10 kV/3 kA (ABB) Prototype

6.5 kV/1.5 kA (Infineon/ABB/Hitachi)

Module IGBT

IGCT

6.5 kV/3.8 kA (ABB)

PP-IGBT

4.5 kV/5 kA (ABB)

4.5 kV/2.4 kA (Hitachi, ABB)

4.5 kV/4.8 kA (Westcode)

1.7 kV/7.2 kA (Mitsubishi)

3.3 kV/3 kA (Infineon/Hitach ABB/Mitsubischi)

V in kV

I in kA

Figure 2.1: Nominal blocking voltage and maximum turn off current ratings of currently commercially available MV high-power devices with turn-off capability

2.2 State of the Art of Medium Voltage Topologies

2.2.1 Topologies, Classification and Application

Nowadays, the importance of medium and high voltage converters is growing due to the increasing demand of high efficient and dynamic converters in industry, traction, power transmission, power distribution and utility applications [1], [2].

High-power medium and high voltage converters are used for different power ranges up to several 100 MW and have found widespread applications in different industry sectors like [18]:

8

- oil and gas for turbo compressors, centrifugal pumps, and reciprocating compressors;
- metal industry with hot and cold rolling mills, sectional steel mill, and blast furnace converter;
- pumps, coal mills, blowers, fans, static var compensator, power converters for wind turbines, and HVDC links;
- mining with mills and conveyors;
- belts, pumps, crushers, blowers, and compressors;
- water pumps and blowers;
- ship propulsion drives, booster-generators, and dredge pumps;
- extruders, pumps, and compressors in chemical, cement, pulp, and paper industry;
- blowers, fans, mixers, and presses.

The general requirements for medium and high voltage converters include characteristics like [18]:
- low costs per power;
- robustness, very high reliability, and availability;
- low line harmonics and low torque ripple;
- easy to integrate, to use, and to maintain;
- low losses;
- minimum space requirements, and low weight.

MV converters can be divided in two categories: direct and indirect converter topologies. Direct voltage converters connect the source and the load directly through the power devices. Indirect converters use an auxiliary energy storage element (inductor or capacitor) called dc-link to connect the grid side and the load side converter. Figure 2.2 shows a classification of the commercially applied topologies in MV applications [4], [2].

Figure 2.3 illustrates two examples of direct converters available on the MV market: the Cycloconverter and the Matrix Converter. The **Cycloconverter (CCV)** [20], [21] is the most widely used direct topology on the MV market with a power range of 1–27 MW. The Cycloconverter is a line commutated converter that directly connects, without

9

intermediate storage, a three-phase electrical network to a three-phase machine. The input voltage has a fixed amplitude and frequency, whereas the fundamental of the output voltage is variable in amplitude and frequency. The fundamental output frequency must not exceed one-third to one-half of the input frequency.

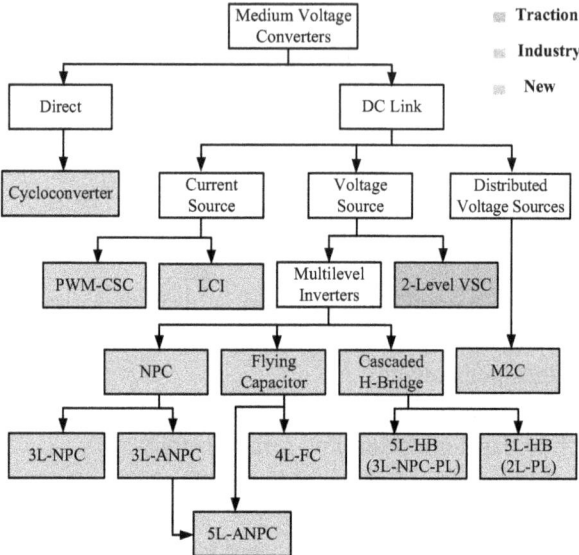

Figure 2.2: Classification of Medium Voltage Converters [4], [19]

The reliability of the converter is very high due to the simple and highly reliable thyristors. Cycloconverters are used in four quadrant operations. Typical applications for the Cycloconverter are rolling mill main drives, mine hoist drives, ship propulsion, and cement mill drives [20].

The **Matrix Converter (MXC)** [22] (see Figure 2.3 (b)) is a direct converter, recently available on the MV drives market. The manufacturer Yaskawa produces Matrix Converter with phase-to-phase voltages from 3.3 to 6.6 kV [23]. The Matrix Converter presents 9 bidirectional switches and an input filter. The converter features low harmonic output currents and power-regeneration function that permits energy saving.

10

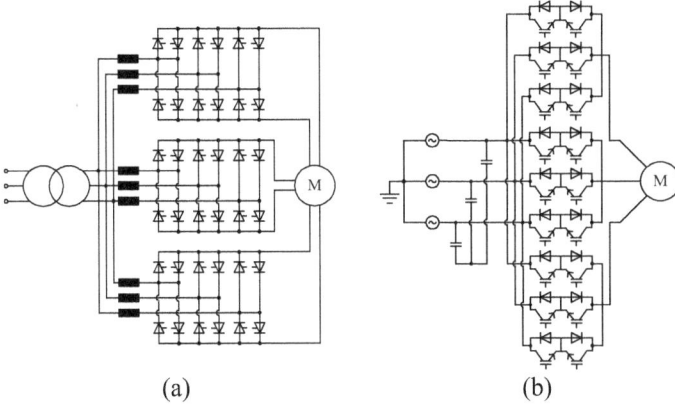

(a)   (b)

Figure 2.3: Direct converter topology: (a) Cycloconverter and (b) Conventional Direct Matrix Converter

AC-AC converters can be further divided into **Current Source Converters (CSC)** and **Voltage Source Converters (VSC),** depending on the energy-storage component of dc-link. Figure 2.4 illustrates a general block diagram of an indirect MV drive. In the case of VSCs, the energy-storage component is a dc-link capacitor to filter the dc-voltage, whereas the load is supplied by variable three-phase voltages. On the other hand, for CSCs, the energy storage component is an inductor to filter the dc-current, whereas the load is supplied by variable three-phase currents. The line- and motor-side LC filters are necessary or optional, depending on the topology and the requirements. A phase-shifting transformer with multiple secondary windings is often employed for the reduction of the line-current distortion. For the VSC rectifier part different topologies can be used: diode bridge rectifiers e.g. 6, 12-, 18-, 24- or 36 pulse rectifiers [24].

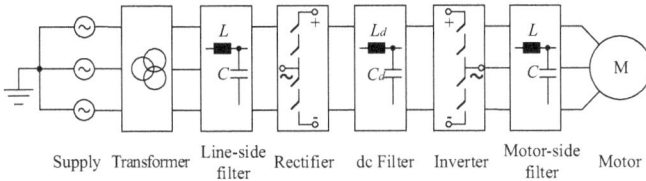

Supply   Transformer   Line-side filter   Rectifier   dc Filter   Inverter   Motor-side filter   Motor

Figure 2.4: General block diagram of a MV converter for drives [2], [24]

11

Two CSC topologies are used in industry: the **Load Commutated Inverter (LCI)** and the **Pulse Width Modulation Current Source Converter (PWM CSC)**. The LCI [21], [25] (see Figure 2.5 (a)) presents a simple converter topology featuring thyristors, low manufacturing costs and high reliability. The LCI is used for two quadrant and four quadrant operation for very large synchronous motor drives with a power rating up to 100 MW. The major drawbacks are the low-input power factor at partial load and considerable current harmonics [26].

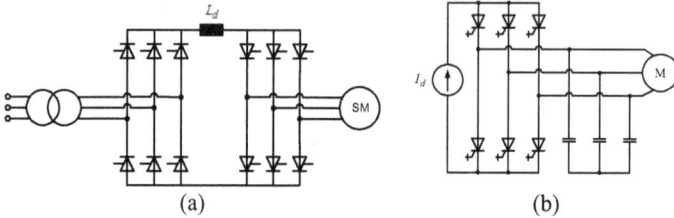

(a)   (b)

Figure 2.5: Current Source Converters: (a) Load Commutated Inverter and (b) Pulse Width Modulation Current Source Converter

The **PWM CSC** [2] (see Figure 2.5 (b)) is designed with turn-off power devices (GTO or IGCT). Pulse Width Modulation is used to generate sinusoidal fundamental output currents, with a switching frequency typically lower than 200 Hz in order to minimize switching losses. The converter has a simple structure and therefore presents a high reliability. However it needs an output filter capacitor bank. The PWM CSC can be used as both load-side (e.g. inverter) and grid-side converter (e.g. rectifier).

The **Two Level-Voltage Source Converter (2L-VSC)** [2] (see Figure 2.6 (a)) is used in medium- and high-power traction and industrial drives, with power devices in a range from 3.3 to 6.5 kV. A direct series connection of switches is used if a higher output voltage is desired. The converter presents a simple structure, a low part count and therefore, a high reliability is achieved. In order to obtain sinusoidal fundamental ac outputs voltages with variable magnitude and frequency, different modulation techniques are used: Pulse Width Modulation (PWM), including third-order harmonic injection, Space Vector Modulation

12

(SVM) and offline optimized pulse patterns, e.g. Selective Harmonic Elimination (SHE).

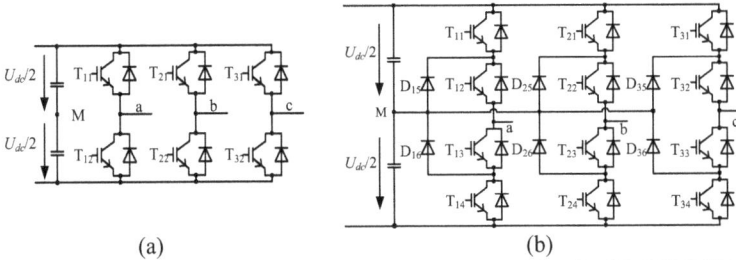

(a)   (b)

Figure 2.6: Voltage Source Converters: (a) 2L-VSC; (b) 3L-NPC VSC

Compared to the 2L-VSC, Multi Level Voltage Source Converters (ML-VSCs) [2] present usually a lower expense of semiconductors for the same installed converter power, a higher expense of capacitive energy storage components and lower costs for a redundant design in medium voltage applications. ML-VSCs present advantages like excellent quality of output voltage and current (e.g. grid converters), high possible frequency of the fundamental component (e.g. High speed drives), and high output voltages (e.g. MV drives with $V_{LL}>4.16$ kV). ML-VSCs can be further divided into three main classes of topologies (see Figure 2.2):

- Neutral Point Clamped Converters, commercialized as Three-Level Neutral Point Clamped Voltage Source Converter (3L-NPC VSC) and Three-Level Active Neutral Point Clamped Voltage Source Converter (3L-ANPC VSC).
- Flying Capacitor Converters, commercialized as Three- and Four-Level Flying Capacitor Voltage Source Converter (3L-FLC VSC, 4L-FLC VSC).
- Five-Level Active Neutral Point Clamped Voltage Source Converter (5L-ANPC VSC), which is a hybrid of a 3L-ANPC VSC and a FLC VSC.
- Cascaded H-Bridge Converters (CHBC), commercialized as Three-Level H-Bridge (3L-HB) and Five-Level H-Bridge (5L-HB) converters.

13

The **3L-NPC VSC** (see Figure 2.6 (b)) consists of 12 active switches connected in parallel with 12 inverse diodes and 6 neutral point clamp diodes. The topology has gained more and more importance because of the good output voltage quality, high reliability and availability. The 3L-NPC VSC presents a superior dynamic behavior by high performance control schemes and an excellent reputation at medium and high power applications [4]. One major disadvantage is the unequal semiconductor temperature distribution, resulting in a limitation of the switching frequency or/and the maximum phase current or semiconductor switch utilization. To overcome this drawback, the 3L-ANPC VSC was introduced. This topology is similar to the 3L-NPC, presenting two extra active switches in parallel with the NPC diodes (D_{x5} and D_{x6}, x = 1, 2, 3) (see Figure 3.1, Chapter 3.1.1).

The **Flying Capacitor VSC** (see Figure 2.7 (a)) presents a modular form, with power sub-modules which consist of two active switches with parallel inverse diodes and a capacitor. Nowadays, only one manufacturer produces 4L-FLC VSCs (see Table 2.4). The advantages of 4L-FLC VSC are the modular structure, the requirement of a standard transformer (in some cases without transformer) and an excellent dynamic behavior [4]. The 4L-FLC VSC is suitable for MV drives with high dynamic requirements, e.g. high frequency of the fundamental output voltage and low harmonic distortions. One disadvantage of this topology is the high expense of flying capacitors at low and medium switching frequency, e.g. f_{lcb}<800-1800 Hz [1], making this topology less attractive than the 3L-NPC VSC for the aforementioned frequency range.

The **5L-ANPC VSC** (see Figure 2.7 (b)) topology is derived from the 3L-ANPC VSC. The voltage levels are increased to 5 by adding a flying capacitor between the series connected switches. The 5L-ANPC has to control the mid-point output and the flying-capacitor voltage. The converter based on the 5L-ANPC topology is used to increase the efficiency of standard, low-cost direct (that means transformerless) drives, overcoming problems such as high harmonic output phase current content, high dv/dt and common mode voltages [27].

14

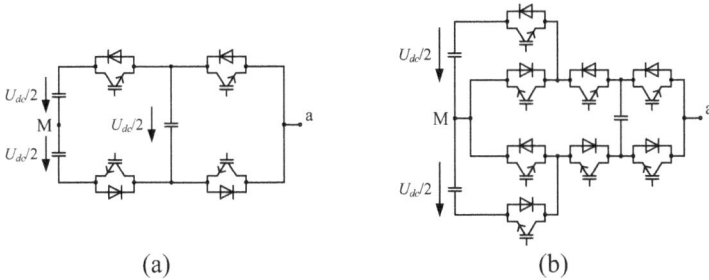

(a)   (b)

Figure 2.7: One phase leg of the (a) 3L-FLC VSC; (b) 5L-ANPC VSC

The 2L-VSC, 3L-NPC VSC, 4L-FLC VSC and 5L-ANPC VSC can be used as both load-side (e.g. inverter) and grid-side converter (e.g. rectifier). The 2L-VSC, 3L-NPC VSC, 3L-FLC VSC, and 4L-FLC VSC have been compared based on costs, efficiency and further characteristics e.g. in [1], [5].

Both 3L-NPC or FLC converters can be extended to generate more output voltage levels, by employing additionally power devices and capacitors. However these extended topologies have not found applications in the industry because of different practical issues like e.g. a higher number of clamping diodes and capacitors. The high number of components implies higher costs. In conventional ML diode clamped VSCs, on the basis of an extension of the 3L-NPC VSC, the control of the partial dc-link voltages is a further problem [27].

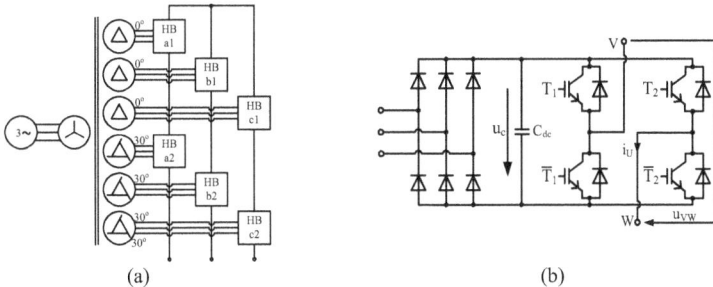

(a)   (b)
**Figure 2.8: Cascaded H-Bridge Converter: (a) 5L-CHB VSC and
(b) 3L-H-Bridge cell**

15

The **Cascaded H-Bridge Converter (CHBC)** (see Figure 2.8 (a)), also known as multicell converter, consists of a series connection of H-bridge power cells (see Figure 2.8 (b)). One power cell consists e.g. of a single-phase 3L-H-Bridge (HB), a dc capacitor, a rectifier and floating transformer windings [24]. The advantage of the multicell converter is the modularity given by the use of 3L-H-Bridge cells. This not only allows a voltage adjustment to the required phase voltage, but also ensures a high availability of the converter by additional redundant cells. In a case of a fault, the failed cell is short-circuited and the converter continues to function [19]. Other advantages of Cascaded H-Bridge Converters are the excellent quality of the grid currents and the output voltage waveforms. Thus, the additional line-side and motor-side filters can be avoided. The drawbacks of this topology are the high number of power devices that imply a high statistical failure rate and the additional costs of the multi-phase transformer that has to be specially designed and manufactured [19], [25].

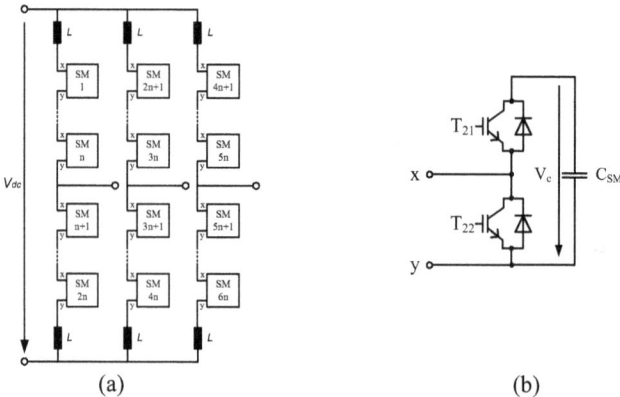

(a)   (b)

Figure 2.9: Modular Multilevel Converter (a) Schematic and
(b) Sub-module in half bridge configuration

The **M2C** also known as MMC (Modular Multilevel Converter) [19], [28] (see Figure 2.9 (a)) also presents a modular structure. Every power cell or sub-module (SM) (see Figure 2.9 (b)) consists of two active switches with anti-parallel diodes and a capacitor. The modular structure of a large number of identical sub-modules presents advantages like: simple realization of redundancy, simple voltage scaling by a series

16

connection of cells and low total harmonic distortion. This permits a filterless configuration for standard machines or grid converters [29]. Thus, the topology is attractive for high speed drives and high voltage dc transmission (HVDC PLUS [29]). However, the M2C presents drawbacks like: higher expense of semiconductors and capacitive energy storage components compared to other Multilevel VSCs (like e.g. 3L-NPC VSC, 5L-ANPC VSC).

Table 2.4 Market overview of selected important manufactures of industrial MV drives [4]

Manufacturer	Type	Power in MVA	Voltage in kV	Topology	Power device	Control
ABB	ACS 1000	0.3–5	2.3; 3.3; 4; 4.16	3L-NPC VSC	IGCT	DTC
	ACS 2000	0.4–1	6.0; 6.9;	5L-NPC VSC	IGBT	DTC
	ACS 5000	1.7–22	(4.16), 6.0; 6.6; 6.9	5L-HB NPC VSC	IGCT	DTC
	ACS 6000	3–27	(2.3); 3; 3.3	3L-NPC VSC	IGCT	DTC
Rockwell Automation	Powerflex 7000	0.15–6.7 (22.5)	2.3; 3.3; 4.1;6.6	PWM CSC	SGCT	VC
GE	VDM5000	1.4–7.2	2.3; 3.3; 4.2	2L-VSC	IGBT	VC
	VDM6000	0.3–8	2.3; 3.3; 4.2	4L-FLC VSC	IGBT	VC
	MV7000	7–9.5	3.3; 6.6	3L-NPC VSC	PP-MV IGBT	VC
Siemens	Sinamics GM150	0.6–10.1	2.3; 3.3; 4.16	3L-NPC VSC	MV IGBT	VC
	Sinamics SM150	5–28	3.3; 4.16	3L-NPC VSC	IGCT/ MV IGBT	VC
	Perfect Harmony	0.3–30	2.3–13.8	ML-SCHB VSC	LV/MV IGBT	VC

Vector Control (VC)/ Direct Torque Control(DTC)

Table 2.4 presents an overview of commercially available converters for industrial MV drives. The most used power devices are IGBT and IGCT, depending on the converter voltage, the output power, the topology, and the commutation voltage. As it can be seen from Table 2.4 the dominant converter topology is the 3L-NPC VSC (e.g. ABB, GE,

17

and Siemens). Only one manufacturer (Rockwell Automation) offers PWM CSCs with symmetrical IGCTs. The Cascaded H-Bridge Converter produced by ABB (5L-HB NPC VSC) consists of an H-bridge of two 3L-NPC phase legs per phase.

2.2.2 Development and characteristics of 3L-NPC VSC

The 3L-NPC VSC topology was proposed in the early 1980s [3], [30]. Since then, it has an increasing demand on the market of MV drives due to the output voltage range (e.g. 2.3kV–4.16 kV), low current and voltage harmonics, low part count, modularity, robustness, high reliability and availability. Nowadays, 3L-NPC VSC using IGBT modules are widely used for pumps, fans, extruders, marine drives and mill drives [4]. Configurations using IGCTs or press-pack IGBTs are employed for hot and cold rolling mill drives, marine drives, compressors, pumps, the Transrapid, and interties [4]. The topology also found first applications in the emerging wind energy market [6], [7]. The 3L-NPC VSC is adapted as a static compensator (STATCOM) for flexible AC transmission systems (FACTS) to compensate the reactive power and/or to control the voltage of power systems [8].

The 3L-NPC VSC is usually applied as load/motor-side converter in two- or four-quadrant operation. The 3L-NPC VSC can be used also as a grid-side converter named active front end (AFE). Usually the load- and grid-side converters are identical 3L-NPC VSCs [6]. Thus, it is possible to regenerate energy back to the utility grid during the braking operation of a motor. Furthermore, an AFE enables an adjustable power factor at the grid-side and sinusoidal grid currents by the use of e.g. field oriented control including optimized pulse patterns. The AFE and the load-side converter can be connected in various configurations to one common dc-bus [6].

Usually, 3L-NPC VSCs are based on Power Electronics Building Blocks (PEBB) [11] which enable a modular converter design. This is an important key for a good maintenance: the modular construction implies

18

fast and effective replacement of components and short repair time. The low part count implies a high reliability and availability. The 3L-NPC VSC based on IGCT PEBBs has been successfully implemented in applications such as [11]:

- "9–27 MVA MV drives for several applications, including rolling mills and marine propulsion;
- 15 MVA back-to-back interties to connect the 50 Hz grid and the 16 2/3 Hz grid of European traction systems;
- 15–60 MVA energy storage systems based on the regenerative fuel cell technology or NiCd battery technology to enhance grid stability or to reduce power fluctuation;
- 22 MVA Dynamic Voltage Restores to safeguard the highly critical processes of a semiconductor plant."

Another important advantage of the topology is an excellent dynamic behavior by high performance control schemes [4]. Depending on the demands of the application, different control schemes can be employed: Direct Torque Control [31], [32], Voltage-Oriented Control, Direct-Power Control [33], and Predictive Control [34], [35]. The most used modulation methods applied in industry are carrier-based PWM [36], [37], [38], Space Vector Modulation [36], [39], [40], [41], and offline optimized pulse patterns (e.g. SHE) [42], [43], [44], [45].

Table 2.5 Power devices, converter voltages and power ranges for 3L-NPC VSC [24]

Power device	Power device blocking voltage in kV	Converter voltage in kV	Converter power in MVA
IGBT Module	3.3, 6.5	2.3, 3.3, 4.16, 6, 6.6	0.8–10
PP-IGBT	4.5	3.3, 3.4	6–32
IGCT	4.5, 5.5	2.3, 3.3, 4.16	0.3–30

Nowadays the 3L-NPC VSCs are manufactured in a power range from 0.8 to 32 MW as shown in Table 2.5. The IGBTs (module or press-pack) present blocking voltages of 3.3, 4.5, and 6.5 kV and IGCTs present blocking voltages of 4.5 and 5.5 kV.

19

Compared to 2L-VSC, the power devices of the 3L-NPC VSC have to block only one-half of the dc-link voltage $U_{dc}/2$. The analysis of the output voltage waveforms of the 3L-NPC VSC reveals also an improved voltage quality compared to 2L-VSC. The steps in the midpoint and line-to-line voltage are reduced by 50% [46]. A disadvantage of the 3L-NPC VSC is that for output voltages between 4.16 and 7.2 kV, series connections of IGBTs or IGCTs are required. The reliability is negatively affected by the high part count in that case. Another issue is the dynamic voltage sharing between the series connected power devices, being solved by the means of active clamping and/or additional RC-snubbers [47], [48], [49]. A 10 kV IGCT [17] was introduced optionally to the series connection of semiconductors for converters in the voltage range from 4.16 to 7.2 kV.

The requirement to control the neutral-point potential is necessary for the safe operation of the converter. Several solutions were presented in the last years [50], [51], [8]. Nowadays, it is considered as a solved problem in the industry. The fault tolerance of the 3L-NPC topology is another safety and reliability problem. Several solutions have been developed and presented (e.g. [52], [53], [54]). Other disadvantages are: the requirement of special configurations for high speed drives, the necessity of a sine filter for standard machines, and the high expense for redundancy [4]. One main drawback of the 3L-NPC VSC is however the unequal loss distribution and the resulting unsymmetrical power device junction temperature distribution.

2.2.3 Conception of the 3L-ANPC VSC

The 3L-ANPC VSC was introduced in 2001 [10] to overcome the unequal loss distribution problem of the 3L-NPC VSC, being patented in 2004 [55]. The topology consists of two extra active switches connected in parallel to the NPC diodes. The active switches enable additional commutation options. Based on additional degrees of freedom, the Active Loss Balancing method improves the unequal junction temperatures distribution of the power semiconductors of one phase leg [56], [57].

20

Furthermore, the influence of different PWM strategies on the loss distribution has been investigated in previously published papers [58], [59].

The 3L-ANPC VSC has gained more interest for high power applications because this converter enables an increase of the maximum converter output power of about 20% with respect to the 3L-NPC VSC [9] at certain operating regimes. A comparative evaluation of 3L-NPC VSC and 3L-ANPC VSC, regarding their maximum output power, efficiency, and power part components, has been investigated in [9], [60]. The 3L-ANPC VSC is especially advantageous in the following applications [61]:

- "High power applications, where the required output power can not be achieved without a serial/parallel connection of power devices.
- MV converters, where the switching frequency should be increased without decreasing the converter power (e.g. applications which require a sine filter or high speed applications).
- MV converters, where the nominal converter current is required at low modulation index and low fundamental frequency (e.g. zero speed operation points, rolling mill application, converter for doubly fed induction generators, etc.)"

The ANPC PEBB [11] based on the new developed 6 kA IGCT has an increased output power capability up to 16 MVA, compared to the 9 MVA-NPC-PEBB. The new ANPC PEBB has been commercialized for pump storage applications [62].

The fault tolerance ability is another important issue concerning safety, availability and reliability of the 3L-ANPC VSC. Control schemes have been proposed to enable continuous operation at device failures [63]. Thus, the reliability and the robustness of the power converter can be improved without additional components [63].

21

2.3 Motivation of the investigation of a New Active Loss Balancing Method for the 3L-ANPC VSC

The junction temperature is a key factor in the design of a converter, because of the direct influence on the converter rating and the lifetime expectancy of the power switches. The thermal analysis of the 3L-NPC VSC has shown an unequal junction temperature distribution in different operating points. In order to ensure that the most stressed power switches do not exceed the maximal allowable junction temperature, the converter output power and switching frequency have to be limited in a conventional 3L-NPC VSC. To overcome this drawback, a junction temperature balance is achieved in the 3L-ANPC VSC applying the Active Loss Balancing (ALB) method [10], [56], [57].

The ALB method represents a decision algorithm that uses the estimated junction temperatures as control variables. The aim of the decision algorithm is to advantageously distribute the switching losses between different groups of semiconductors and thus, achieving an equal distribution of the junction temperatures. The highest junction temperature is substantially reduced, enabling an increase of the converter output power and/or switching frequency [9].

In this work a new balancing method is proposed to improve the performance of the 3L-ANPC VSC. The aim of the balancing method is to reduce the highest junction temperature. Similar to the ALB method, the new proposed method uses the different commutation possibilities given by the active switches to clamp the neutral tap of the converter. The new balancing method predicts the thermal behavior of the converter to advantageously distribute the switching and the conduction losses. The performance of the Predictive Active Loss Balancing (PALB) method is evaluated in comparison to the ALB method.

22

3 Three-Level Neutral Point Clamped Voltage Source Converters

The 3L-NPC VSC is described in the first part of this chapter. The converter structure, switch states and commutations are presented, followed by the employed pulse width modulation schemes. The analysis of the loss and junction temperature distribution shows substantial limitations of this topology and provides the motivation for this work.

The second part of this chapter describes the 3L-ANPC VSC. The main features like the switch states and the commutations within this topology are presented. The principles of a loss balancing method applied to the 3L-ANPC VSC are explained, followed by the effect on the loss and junction temperature distribution compared to the 3L-NPC VSC.

3.1 The 3L-NPC VSC

3.1.1 Structure of the 3L-NPC VSC

Figure 3.1 presents the simplified topology of the 3L-NPC VSC. The converter consists of three phases joined at the common dc-link capacitors. The 3L-NPC VSC contains 12 active switches connected in parallel with 12 inverse diodes and 6 neutral point clamp diodes. The switches T_{11}, T_{21}, T_{31} and T_{14}, T_{24}, T_{34} are designated as outer switches (T_{out}) and their inverse diodes are designated as outer diodes (D_{out}). The switches T_{12}, T_{22}, T_{32} and T_{13}, T_{23}, T_{33} are designated as inner switches (T_{in}), while their parallel inverse diodes are designated as inner diodes (D_{in}). The diodes D_{15}, D_{16}, D_{25}, D_{26}, D_{35} and D_{36} connected to the neutral point are the NPC diodes and are designated as D_{NPC}.

23

The dc-bus capacitor on the dc-side of the converter consists of two identical series-connected capacitors, providing a neutral point "M". The voltage of each dc capacitor is the half of the dc voltage, $U_{dc}/2$. The dc-link capacitors are charged and discharged by the neutral current i_M. Different solutions regarding the balancing of the dc-link capacitors were presented in [8], [50], [51]. In the analysis of the converter, the dc-bus capacitors are considered to be sufficiently large to ensure a constant dc-voltage $U_{dc}/2$.

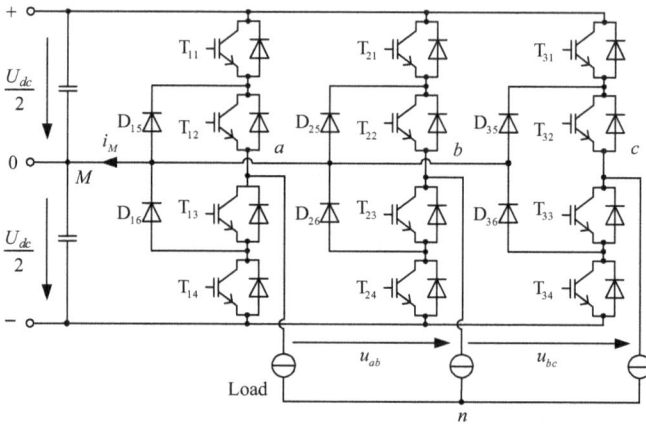

Figure 3.1: Three-Level Neutral Point Clamped Voltage Source Converter [9]

In the case of the 3L-NPC VSC featuring IGCTs, a clamp circuit is required. The clamp circuit has the following tasks: to limit the di/dt during the IGCT turn-on and the diode turn-off, to keep the overvoltage during the turn-off below the IGCT breakdown voltage, to maintain the commutation time within a few percent of the pulse width and to limit the short circuit current. The clamp circuit reduces the turn-on losses from the power semiconductors, transferring them from the IGCTs to the resistive components of the clamp circuit. The clamp circuit also increases the part count and the converter complexity [9].

3. Three-Level Neutral Point Clamped Voltage Source Converter

3.1.2 3L-NPC VSC switch states and commutations

There are three dc-terminals: the positive dc-rail symbolized with "+", the neutral point "0" and the negative dc-rail "−". Each phase node (a, b, c) can be connected to each dc-terminal ($-U_{dc}/2$, 0, $U_{dc}/2$). There are 27 possible switch states being calculated according to

$$n_{sw} = n^p = 3^3 = 27 \qquad (3.1)$$

where n is the number of dc-link voltage levels and p the number of phases.

Table 3.1 Definition of switch states for the 3L-NPC VSC

Switch state	Switch positions				Phase-to-neutral voltage
	T_{x1}	T_{x2}	T_{x3}	T_{x4}	
+	1	1	0	0	$+U_{dc}/2$
0	0	1	1	0	0
−	0	0	1	1	$-U_{dc}/2$

Table 3.1 presents the 3L-NPC VSC switch states per phase. "1" designates the on-state and "0" designates the off-state of the switch. Figure 3.2 presents the current paths for a positive and negative phase current during the possible switch states. The phase current is considered positive if it flows from the dc-side to the load side and negative vice versa.

The "+" state is obtained by turning on T_{x1} and T_{x2}, while T_{x3} and T_{x4} are in the off-state. In this case, the phase-to-neutral point voltage u_{xM} is $U_{dc}/2$. In the "+" state, the path for positive current is provided by the upper active switches T_{x1} and T_{x2}, while the path for negative current is provided by the upper diodes D_{x1} and D_{x2} (see Figure 3.2 (a)).

The "0" state is achieved when the inner devices T_{x2} and T_{x3} are turned on. The phase-to-neutral point voltage is clamped to the neutral point "M". In this case, the direction of the load current determines whether the upper or lower path of the neutral tap is used. The upper path of the neutral tap is given by D_{x5} and T_{x2} and the lower path by D_{x6} and T_{x3}. The inner power devices T_{x2} and T_{x3} must always be in the on-state to provide an open path in case that the current reverses the direction (see Figure 3.2 (b)).

The "–" state is accomplished by turning on the lower two switches (T_{x3} and T_{x4}), leading to a phase-to-neutral point voltage of – $U_{dc}/2$. In the "–" state, the negative current flows through T_{x3} and T_{x4}, whereas the positive current flows through the two lower diodes D_{x3} and D_{x4} (see Figure 3.2 (c)).

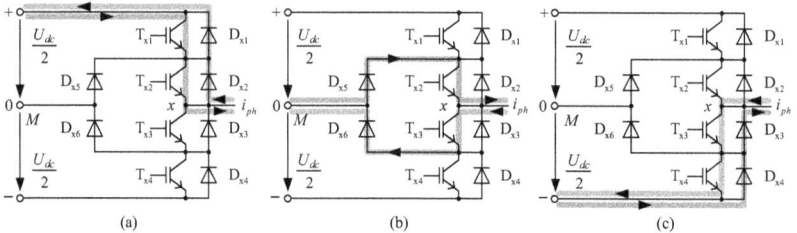

(a)   (b)   (c)

Figure 3.2: Conduction paths for the 3L-NPC VSC [9]:
(a) "+" state; (b) "0"state; (c) "–" state

The conduction losses occur in the power semiconductors that are carrying the load current (see Figure 3.2). The distribution of the conduction losses is summarized in Table 3.3.

Table 3.2 Characteristics of forced and natural commutations

Commutation type	Characteristics	Conditions
Forced (capacitive)	negative gradient of the output power; initiated by an active turn-off transient;	if $i_{ph}>0$ and ("+"→"0" or "0"→"–") if $i_{ph}<0$ and ("0"→"+" or "–"→"0")
Natural (inductive)	positive gradient of the output power; initiated by an active turn-on transient;	if $i_{ph}>0$ and ("0"→"+" or "–"→"0") if $i_{ph}<0$ and ("+"→"0" or "0"→"–")

26

The switching losses occur during the commutation process between different switch states. Two types of commutations can be classified according to the gradient of the converter output power: forced (capacitive) and natural (inductive) commutations [65] (see Table 3.2). During the capacitive commutation, the active switches experience turn-off losses. During the inductive commutation, the active switches present turn-on losses and the diodes present turn-off losses. The diode turn-on losses are insignificant compared to the aforementioned types of losses and therefore, they are neglected [9], [65].

There are eight different commutations in the 3L-NPC VSC: "+"↔"0" and "0"↔"−" for positive and negative phase current. Figure 3.3 illustrates the commutations and switching losses in the 3L-NPC VSC. The current path of the switching active devices is marked with a blue line and the current path of the switching passive device is marked with a green line. The power semiconductors that present switching losses are marked with a hatched circle.

Figure 3.3 (a) shows the commutation "+"↔"0" for a positive phase current. The commutation "+"→"0" is initiated by the turn-off of T_{x1}. In the following step, T_{x3} is turned on after a dead time to ensure that T_{x1} has been completely turned off. Thus, the current is forced to commutate from the path formed by T_{x1} and T_{x2} to the upper neutral tap formed by D_{x5} and T_{x2}. The clamping diode D_{x5} is positive biased by the positive phase current. In this case, T_{x1} experiences turn-off losses. Although T_{x3} is turned on, it does not experience turn-on losses since it does not carry any current after the commutation. In order to go from "0" state back to "+" state ("0"→"+"), all the switching transitions take place in the reversed order: T_{x3} is turned off first and after a dead time, T_{x1} is turned on. T_{x1} experiences turn-on losses and D_{x5} experiences recovery losses. Figure 3.3 (b) illustrates the commutation "0"↔"−" for a positive phase current. The commutation "0"→"−" is initiated by the turn-off of T_{x2}. Since T_{x3} remains in the on-state from the zero state, only T_{x4} is turned on after a dead time. The active switch T_{x2} experiences turn-off losses. For the reverse commutation "−"→"0", all switching transitions take place in the reversed order: first, T_{x4} is turned off and after a dead time, T_{x2} is turned on. Thus, the phase current commutates

27

from the negative dc-rail formed by D_{x3} and D_{x4} to the upper neutral tap formed by D_{x5} and T_{x2}. The diode D_{x4} experiences recovery losses and T_{x2} experiences turn-on losses. The commutations for negative current depicted in Figure 3.3 (c) and (d) occur in a similar manner. The switching loss allocation is summarized in Table 3.4.

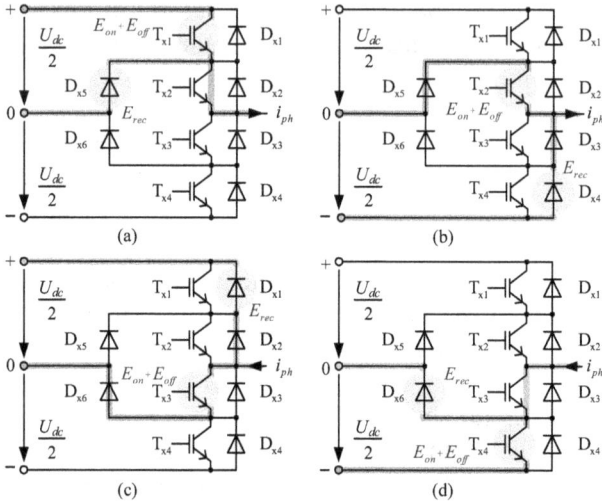

Figure 3.3 Commutations and switching losses in the 3L-NPC VSC [9]
Positive phase current: (a) "+"↔"0" and (b) "0"↔"–";
Negative phase current: (c) "+"↔"0" and (d) "0"↔"–"

Table 3.3 Conduction losses in the 3L-NPC VSC [9]

	T_{x1}	D_{x1}	T_{x2}	D_{x2}	T_{x3}	D_{x3}	T_{x4}	D_{x4}	D_{x5}	D_{x6}
	Positive phase current									
+	×		×							
0			×						×	
–					×		×			
	Negative phase current									
+	×		×							
0					×					×
–					×		×			

28

Table 3.4 Switching losses distribution in the 3L-NPC VSC [9]

	T_{x1}	D_{x1}	T_{x2}	D_{x2}	T_{x3}	D_{x3}	T_{x4}	D_{x4}	D_{x5}	D_{x6}
Positive phase current										
"+"↔"0"	×								×	
"0"↔"–"			×					×		
Negative phase current										
"+"↔"0"		×			×					
"0"↔"–"							×			×

3.1.3 Modulation of the 3L-NPC VSC

The modulation is an important factor that influences the performance of the converter because it determines the spectral content of the output voltage waveform. Another important aspect is the influence of the modulation on the losses in the converter and their distribution among the power devices [9]. The modulation strategy is characterized by the pulse width, the pulse position within the carrier or the half carrier period and the pulse sequence ([9], [66]). The interdependence of these three factors determines the harmonic performance and the maximum output voltage of a converter [66]. There are two commonly known PWM techniques: carrier-based modulation and Space Vector Modulation (SVM). The carrier-based modulation is easier to implement, while the SVM presents a better understanding [67]. For lower switching frequencies in MV converters, optimized pulse patterns like e.g. Selective Harmonic Elimination (SHE) [2], [43] is an attractive method because of the improved harmonic content.

3.1.3.1 The carrier-based PWM and SVM

The carrier-based PWM method is a per-phase modulation technique. The carrier-based modulation compares a reference wave form (e.g. sinusoidal) with a triangular carrier to determine the switching instants. There are two basic possible sampling techniques for PWM: natural sampled PWM and regular sampled PWM. The natural sampled

PWM is suitable for an analogous implementation rather than for digital implementation. In the case of regular sampling, the pulse widths have an analytical form easy to calculate and thus, this technique is suitable for a digital implementation. Regular sampling can be either symmetrical, where the reference is sampled every carrier interval or asymmetrical, where the reference is sampled every half carrier interval. The asymmetrical regular sampling presents a better harmonic performance compared to the symmetrical sampling [64].

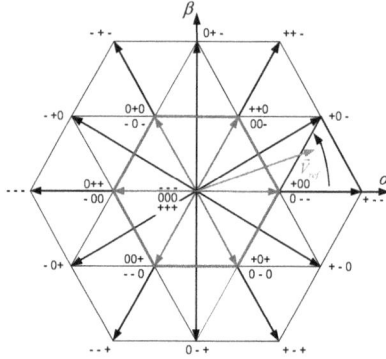

Figure 3.4: Space vector diagram of the 3L-NPC VSC

The modulation using SVM is a three-phase approach. Figure 3.4 illustrates the space vector diagram for the 3L-NPC VSC. The converter presents 27 switch states corresponding to all the combinations of connections of each phase to the dc-link terminals "+", "0" and "–". There are 19 distinguishable space vectors in the α-β-reference plane frame corresponding to the different switch states. The reference phasor $\vec{V}_{ref}$ rotates in the α-β-reference plane frame (Clarke transformation) with the converter fundamental frequency. The length of the phasor determines the converter output voltage. The phasor is sampled within each switching period and is formed by the Nearest Three space Vectors (NTV). By using the NTV method, the voltage harmonics and the current ripple are minimized [66].

Although SVM and regular sampled carrier based PWM present different implementation algorithms, the two methods are both regular sampled systems. The equivalence between the two methods is shown in [9], [66]. The main advantage of the SVM is the explicit identification of the pulse placements. This can be seen as a degree of freedom that can be used to improve the harmonic distortion. In the case of carrier-based PWM, the manipulation of pulse position is achieved by the addition of a common-mode offset to the three-phase references. The common-mode offsets in the converter output voltages are canceled in the phase voltages of the load [9] if the neutral-point of the load is isolated.

By using a pure sinusoidal reference, the maximum linear modulation index is limited to $m_a=1$ [66]. In electric drive systems with isolated neutral-points of the loads, the voltage utilization level can be improved by adding a third harmonic with one-sixth of the fundamental amplitude. Thus, the modulation index (also called modulation depth) can be increased to its theoretical maximum with a linear modulation of $m_a=1.1555$ [66] and the harmonic distortion can be reduced. An even better harmonic performance is obtained by adding a third harmonic with one-quarter of the fundamental amplitude. However, the linear modulation range is limited to $m_a=1.12$ [9].

The carrier-based PWM using sinusoidal references with and without third harmonics can readily be implemented for 2L-VSCs and 3L-NPC VSCs. However, a better harmonic spectrum can be achieved by exact centering of the active pulses in each sample period [66]. This method has become industry practice and will be referred as SVM throughout the thesis. To generate balanced three-phase output voltages in the 3L-NPC VSC, two triangle carrier waveforms are compared to the three-phase reference signals. The two triangular waveforms have identical shape and phase (see Figure 3.7 (a)). This method is called Phase Disposition (PD) modulation [66]. The PD method, using asymmetrical regular sampling, was identified to be an advantageous modulation for the 3L-NPC VSC [9].

The three-phase sinusoidal reference signals are defined as:

31

$$u_{ref,a} = \hat{U}_{con,1} \cdot \sin(2\pi f_1 \cdot t)$$

$$u_{ref,b} = \hat{U}_{con,1} \cdot \sin(2\pi f_1 \cdot t - \frac{2}{3}\pi)$$ (3.2)

$$u_{ref,c} = \hat{U}_{con,1} \cdot \sin(2\pi f_1 \cdot t - \frac{4}{3}\pi)$$

where f_1 is the fundamental output frequency of the converter and $\hat{U}_{con,1}$ is the amplitude of the fundamental component of the reference signals.

The frequency modulation ratio m_f is calculated as:

$$m_f = \frac{f_{sw}}{f_1},$$ (3.3)

where f_{sw} is the frequency of the upper carrier signal $u_{tri,up}$ and the lower carrier signal $u_{tri,low}$.

The modulation index m_a is defined as:

$$m_a = \frac{\hat{U}_{con,1}}{\hat{U}_{tri}} = \frac{\hat{u}_{xM,1}}{U_{dc}/2},$$ (3.4)

where the amplitude $\hat{U}_{tri}$ of the triangular signal is kept constant at $U_{dc}/2$ and $\hat{U}_{con,1}$ is the amplitude of the fundamental component of the modulation signals. For the aforementioned three sinusoidal reference signals, the common-mode offset function for two-level converter is defined as:

$$u_{off} = -\frac{u_{max} + u_{min}}{2},$$ (3.5)

with

$$u_{max} = \begin{cases} u_{ref,a} & \text{if} \quad u_{ref,a} \geq u_{ref,b}, u_{ref,c} \\ u_{ref,b} & \text{if} \quad u_{ref,b} \geq u_{ref,a}, u_{ref,c} \\ u_{ref,c} & \text{if} \quad u_{ref,c} \geq u_{ref,a}, u_{ref,b} \end{cases} \text{ and } u_{min} = \begin{cases} u_{ref,a} & \text{if} \quad u_{ref,a} \leq u_{ref,b}, u_{ref,c} \\ u_{ref,b} & \text{if} \quad u_{ref,b} \leq u_{ref,a}, u_{ref,c} \\ u_{ref,c} & \text{if} \quad u_{ref,c} \leq u_{ref,a}, u_{ref,b} \end{cases}$$

The reference waveform $u_{2L\text{-}SVM}$ for phase a (see Figure 3.5 (a)) is calculated as:

$$u_{2L\text{-}SVM} = u_{ref,a} + u_{off} .$$ (3.6)

The application of the resulting reference $u_{2L\text{-}SVM}$ to PD modulation of the 3L-NPC VSC does not yield centered active vectors within each half-switch period [68], [69]. In order to overcome this issue, the modulating signals are shifted into a common carrier band.

$$u_x^{'} = \left(u_x + u_{off} + \frac{U_{dc}}{2} \right) \text{mod} \left(\frac{U_{dc}}{2} \right) - \frac{U_{dc}}{4} \qquad \left(x = ref,a;\ ref,b;\ ref,c \right),$$ (3.7)

where (a mod b) delivers the remainder of the division (a/b). The offset equation (3.5) is then applied to the modified references

$$u_{off}^{'} = -\frac{u_{min}^{'} + u_{max}^{'}}{2},$$ (3.8)

with

$$u_{max}^{'} = \begin{cases} u_{ref,a}^{'} & if & u_{ref,a}^{'} \geq u_{ref,b}^{'}, u_{ref,c}^{'} \\ u_{ref,b}^{'} & if & u_{ref,b}^{'} \geq u_{ref,a}^{'}, u_{ref,c}^{'} \\ u_{ref,b}^{'} & if & u_{ref,c}^{'} \geq u_{ref,a}^{'}, u_{ref,b}^{'} \end{cases} \quad and \quad u_{min}^{'} = \begin{cases} u_{ref,a}^{'} & if & u_{ref,a}^{'} \leq u_{ref,b}^{'}, u_{ref,c}^{'} \\ u_{ref,b}^{'} & if & u_{ref,b}^{'} \leq u_{ref,a}^{'}, u_{ref,c}^{'} \\ u_{ref,b}^{'} & if & u_{ref,c}^{'} \leq u_{ref,a}^{'}, u_{ref,b}^{'} \end{cases}$$

The final modified reference signals are calculated as:

$$u_{3L\text{-}SVM} = u_x + u_{off} + u_{off}^{'} \qquad \left(x = ref,a;\ ref,b;\ ref,c \right).$$ (3.9)

Figure 3.5 (b) illustrates the resulting reference waveforms. For $m_a=1.155$, the 3L-SVM waveform takes the shape of the 2L-SVM waveform [9]. For a small modulation index $m_a \leq 0.5775$, the 2L-SVM in one carrier band [9] was identified to be especially advantageous regarding to the harmonic performance, loss distribution, and the neutral-point potential control. Applying this modulation technique, the 3L-NPC VSC behaves similar to a 2L-VSC, using the upper ("0" to "+") or the lower set ("–" to "0") of redundant switch states from the inner hexagon (see Figure 3.4). In order to obtain the 2L-SVM in a certain time interval in one carrier band, the reference waveforms for a 2L-SVM are shifted into the upper or lower carrier band by a common-mode dc-offset of ±0.5, respectively. The dc-offset can be arbitrarily switched between +0.5 and −0.5. The alternation of the dc-offset is used to distribute the losses between the converter switches and to control the neutral-point potential. The dc-offset is equally added to three-phase sinusoidal reference signals and is canceled in the line-to-line voltage. Therefore, the use of the dc-offset has no influence on the phase voltage of the load. Figure 3.6 depicts the equivalent reference voltage waveform for a 2L-SVM of a 3L-NPC VSC. Particularly for this case, the dc-offset changes its polarity with the half of the fundamental frequency.

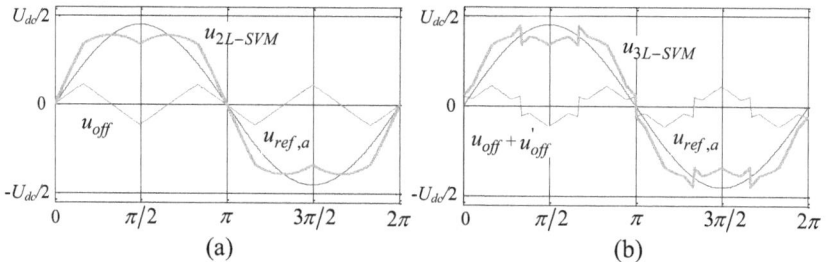

Figure 3.5 Reference waveforms for carrier-based PWM (m_a=0.9)
(a) SVM for two-level converter (2L-SVM); (b) three-level SVM (3L-SVM)

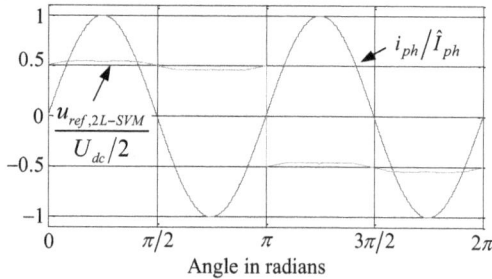

Figure 3.6 Reference waveform for 2L-SVM of a 3L-NPC VSC (m_a=0.05)
and normalized phase current

3.1.3.2 Voltage waveforms

The active switches of the 3L-NPC VSC are controlled based on the comparison of the carrier signals and the modified sinusoidal reference signals. The following output voltages result:

$$u_{xM} = \begin{cases} \dfrac{U_{dc}}{2} & \text{if } u_{ref,x} > u_{tri,up} \\ 0 & \text{if } u_{tri,down} < u_{ref,x} < u_{tri,up} \\ -\dfrac{U_{dc}}{2} & \text{if } u_{ref,x} < u_{tri,down} \end{cases} \qquad (3.10)$$

$$x = a,b,c$$

Figure 3.7 shows the resulting voltage waveforms of the 3L-NPC VSC for the 3L-SVM. Figure 3.7 (a) presents the two triangular carrier signals for a PD modulation and the reference waveform for phase *"a"*. The phase-midpoint output voltage u_{aM} (see Figure 3.7 (b)) alternates between three values: $-U_{dc}/2$, 0, and $U_{dc}/2$. The maximum line-to-line voltage U_{dc} is generated by connecting two phases to the opposite dc-rail. The voltage waveform comprises five voltage levels: $-U_{dc}$, $-U_{dc}/2$, 0, $U_{dc}/2$, and U_{dc}. The intermediate voltage levels $U_{dc}/2$ and $-U_{dc}/2$ are generated by the connection of one phase to the neutral point "0" and the other phase to the positive dc-rail "+" or to the negative dc-rail "−", respectively.

The intermediate voltage levels $U_{dc}/2$ and $-U_{dc}/2$ are generated by the connection of one phase to the neutral point "0" and the other phase to the positive dc-rail "+" or to the negative dc-rail "−", respectively. Figure 3.7 (c) shows the resulting line-to-line voltage calculated as:

$$u_{ab} = u_{aM} - u_{bM}. \qquad (3.11)$$

The common mode voltage shown in Figure 3.7 (d) presents five voltage levels: $-U_{dc}/3$, $-U_{dc}/6$, 0, $U_{dc}/6$, and $U_{dc}/3$ and is calculated as:

$$u_{nM} = \frac{1}{3}\left(u_{aM} + u_{bM} + u_{cM}\right). \qquad (3.12)$$

3. Three-Level Neutral Point Clamped Voltage Source Converter

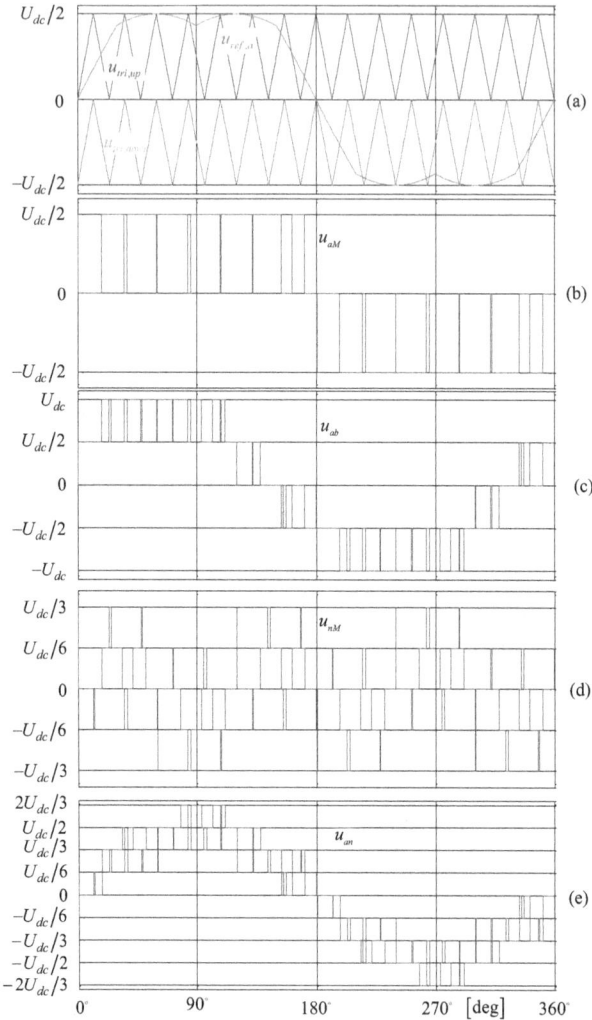

Figure 3.7 Voltage waveforms of the 3L-NPC VSC: (a) voltage reference for phase "a" $u_{ref,a}$ and triangle carrier signals $u_{tri,up}$ and $u_{tri,low}$; (b) phase-midpoint output voltage u_{aM}; (c) line-to-line voltage u_{ab}; (d) common-mode voltage u_{nM}; (e) line to neutral point of the load u_{an} (m_a=1.15, 3L-SVM, f_{sw}=750 Hz, f_1=50 Hz)

36

The line to neutral voltage of the load shown in Figure 3.7 (e) presents 9 voltage levels: $\pm 2U_{dc}/3$, $\pm U_{dc}/2$, $\pm U_{dc}/3$, $\pm U_{dc}/6$, and 0 and is calculated as:

$$u_{an} = u_{aM} - u_{nM} \, . \tag{3.13}$$

3.1.3.3 Selective Harmonic Elimination PWM

MV drives are usually operated at low switching frequency in order to reduce the semiconductor switching losses. By using carrier-based PWM or SVM at frequency modulation ratios below $m_f \leq 15$, the low-order harmonics become more significant [70]. A better harmonic performance at low switching frequency is achieved by optimized pulse patterns. SHE-PWM is one realization option for optimized pulse patterns, which is often used in grid side converters [42], [44], [45]. Thus, the harmonic filter can be minimized or even avoided.

The principle of SHE is based on the Fourier analysis of the phase-midpoint voltage u_{xM} [43]:

$$u_{xM}(\omega \cdot t) = \sum_{h=1}^{\infty} \hat{u}_{xM,h} \cdot \sin(h \cdot \omega \cdot t) \, , \tag{3.14}$$

$$\hat{u}_{xM,h} = \frac{4}{\pi} \cdot \frac{U_{dc}}{2} \cdot \frac{1}{h} \cdot \left[\sum_{k=1}^{N} (-1)^{k+1} \cdot \cos\left(h \cdot \alpha_k\right) \right] \quad x = a,b,c \, , \tag{3.15}$$

where $\hat{u}_{xM,h}$ is the peak voltage of the harmonic h component (e.g. 5th, 7th, 11th, 13th etc.), u_{xM} is the instantaneous phase voltage, N is the number of switching angles per quarter-fundamental period, and α_k are the switching angles. The switching angles must satisfy the following condition:

$$\alpha_1 \leq \alpha_2 \leq ... \leq \alpha_N \leq \pi/2 \, . \tag{3.16}$$

37

Certain lower harmonic components are eliminated and the requested fundamental voltage $\hat{u}_{xM,1}$ is obtained if the proper switching angles are determined. Because the lowest harmonic components are the most undesirable, usually they are chosen to be eliminated. The higher harmonics can be attenuated using filter circuits. The triple harmonics are absent in a three-phase system with isolated neutral-point. The even harmonics are suppressed by simply maintaining the symmetry of the voltage u_{xM} (quarter-wave symmetry) [43].

The SHE modulation is characterized by a defined number of angles per quarter-period. The SHE-3α presents three angles per quarter-period, whereas SHE-5α presents five angles per quarter-period. For SHE-3α, the 5th and 7th harmonics are eliminated and in the case of SHE-5α, the 5th, 7th, 11th, 13th harmonics are eliminated. Figure 3.8 shows the switching pattern for SHE-3α and SHE-5α for m_a=1.15 and pf=1.

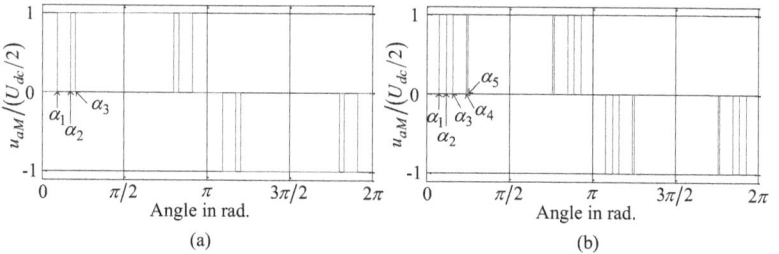

Figure 3.8 Switching patterns for m_a=1.15, pf=1:
(a) SHE-3α and (b) SHE-5α

As an example, the switching angles for SHE-3α are calculated. In order to find the proper switching angles, the system described by the equation (3.17) has to be solved. The system (3.17) represents a set of nonlinear and transcendental equations that can be solved with the aid of a numerical algorithm, e.g. Newton-Rapson method [2]. For SHE-5α, the switching angles can be calculated in a similar manner.

$$\hat{u}_{aM,h=1} = m_a \cdot \frac{U_{dc}}{2} \Rightarrow \cos(\alpha_1) - \cos(\alpha_2) + \cos(\alpha_3) = m_a \cdot \frac{\pi}{2} \qquad (3.17)$$

$$\hat{u}_{aM,h=5} = 0 \Rightarrow \cos(5 \cdot \alpha_1) - \cos(5 \cdot \alpha_2) + \cos(5 \cdot \alpha_3) = 0$$

$$\hat{u}_{aM,h=7} = 0 \Rightarrow \cos(7 \cdot \alpha_1) - \cos(7 \cdot \alpha_2) + \cos(7 \cdot \alpha_3) = 0.$$

3.1.4 Loss and temperature distribution in the 3L-NPC VSC

The loss distribution in the 3L-NPC VSC depends on the power factor (pf), the modulation index (m_a) and the modulation scheme. Four operating points are identified as the most critical, regarding the unequal distribution of the losses. These operating points are located at the boundary of the converter operation range: maximal and minimal modulation index, m_a=1.15 and $m_a \approx 0$, respectively, at power factors of pf=1 (e.g. inverter operation) and pf=-1 (e.g. rectifier operation) [9], [56]. In each critical operating point, one group of semiconductors is stressed with a significant amount of losses compared to the other groups of semiconductors. Thus, the loss and temperature distribution of the 3L-NPC VSC is unbalanced. The maximal junction temperature of the most stressed group of semiconductors limits the switching frequency and the maximum phase current of the converter. All operating points in-between are less critical.

Table 3.5 shows the critical operation points of the 3L-NPC VSC. Since the 2L-SVM was identified to be especially advantageous for a small modulation index $m_a \leq 0.5775$ [41], the loss and temperature distribution in the 3L-NPC VSC is analyzed for case C and case D for both 2L-SVM and 3L-SVM.

Table 3.5 Critical operation points of the 3L-NPC VSC

	Power factor	Modulation index	Modulation scheme
Case A	pf=1	m_a=1.15	3L-SVM
Case B	pf=-1	m_a=1.15	3L-SVM
Case C	pf=1	m_a=0.05	3L-SVM, 2L-SVM
Case D	pf=-1	m_a=0.05	3L-SVM, 2L-SVM

On the base of an example, the loss and junction temperature distribution associated with each critical operation point is analyzed for a 3.3 kV 3L-NPC VSC featuring 4.5 kV PP IGBTs and 4.5 kV PP diodes. The simulation parameters are summarized in Table 3.6. The thermal model, the loss and junction temperature calculation methods are described in detail in Chapter 4 and Section 5.2.

Table 3.6 Simulation parameters of the 4.5 kV PP IGBT and 4.5 kV PP diode

Parameter	Value	Description
IGBT Westcode	T1200EB45E	PP IGBT (U_{CE}=4.5 kV, $I_{C,n}$=1.2 kA)
Diode Infineon	D1031SH45T	PP Diode (U_{AK}=4.5 kV, $I_{F,n}$=2.3 kA)
U_{dc}	5000 V	dc voltage
$i_{ph,rms}$	1200 A	nominal phase current
f_{sw}	450 Hz	switching frequency
f_1	50 Hz	fundamental frequency
ϑ_a	50 °C	ambient temperature
$\vartheta_{j,max}$	125 °C	maximal junction temperature

Figure 3.9 shows the loss distribution in the 3L-NPC VSC for the critical operating points. Figure 3.9 (a) illustrates case A, where T_{x2} presents substantial conduction losses because the switch conducts the phase current for the entire positive half wave. T_{x1} conducts for nearly the same time as T_{x2} because the modulation index is high. Thus, T_{x1} and T_{x2} experience almost the same amount of conduction losses. Since D_{x5} conducts for short duty cycles during "0" state, the diode experiences small conduction losses. T_{x1} commutates with D_{x5}, producing switching losses. For a negative phase current, power losses occur in a similar manner in T_{x4}, T_{x3} and D_{x6}. The outer switches T_{out} are the most stressed devices.

Figure 3.9 (b) shows the case B, where D_{x4} and D_{x3} experience significant conduction losses because they conduct the phase current for almost the entire positive wave. T_{x2} presents small conduction losses because the switch conducts for short duty cycles. The commutations between the "0" and "–" states cause switching losses in T_{x2} and D_{x4}. The power devices D_{x1}, D_{x2}, and T_{x3} present a similar loss distribution for a negative phase current. For case B, the outer diodes D_{out} are the most stressed power semiconductors.

40

Figure 3.9 (c) illustrates the case C for two modulation types: 3L-SVM and 2L-SVM. For 3L-SVM, the phase leg will be in the "0" state during most of the time. D_{x5} is stressed with a significant amount of conduction losses because the diode conducts the phase current for almost the entire positive wave. Similar to case A, T_{x2} presents conduction losses. The distribution of switching losses remains unchanged compared to case A. Power losses occur in a similar way in T_{x4}, T_{x3} and D_{x6} for a negative phase current. D_{NPC} and T_{in} are the most stressed devices.

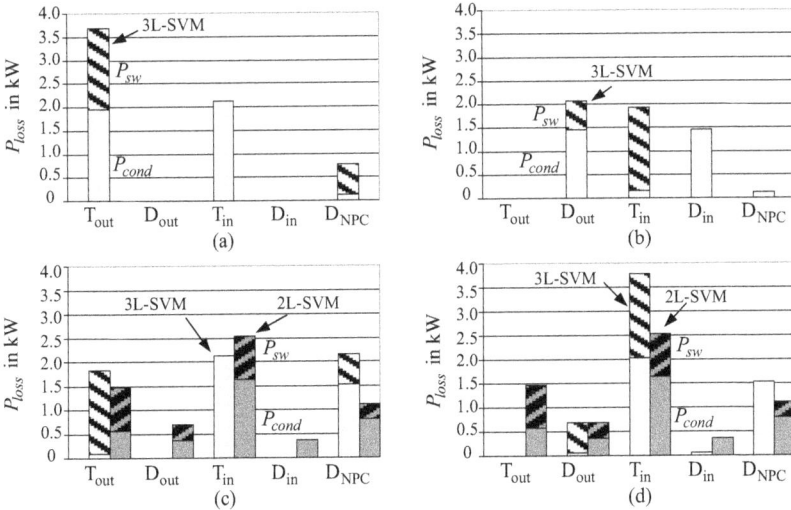

Figure 3.9 Loss distribution in the 3.3 kV 3L-NPC VSC using 4.5 kV PP Westcode T1200EB45E IGBTs and 4.5 kV PP Infineon D1031SH45T diodes; U_{dc}=5000 V, $i_{ph,rms}$=1200 A, f_{sw}=450 Hz, f_1=50 Hz, ϑ_a = 50°C ; (a) pf=1, m_a=1.15; (b) pf=-1, m_a=1.15; (c) pf=1, m_a=0.05; (d) pf=-1, m_a=0.05

Figure 3.9 (d) shows the case D using 3L-SVM and 2L-SVM. For 3L-SVM, D_{x5} and T_{x2} conduct the positive phase current most of the time. Thus, they are stressed with significant conduction losses. The switching loss distribution is identical to case B. The most stressed power devices are the inner active switches T_{in}. When the 2L-SVM is

41

applied, the loss distribution in case C and case D are similar. The similar loss distribution may be explained as follows: for small modulation index $m_a \approx 0$, the reference signal approaches the shape of the common-mode dc-offset $|u_{2L-SVM}| \to 0.5$. Thus, the duty cycle remains constant over time at $d \approx 0.5$. The converter stays for an equal amount of time in the "+" and "0" states when $u_{2L-SVM} \approx 0.5$, and in the "–" and "0" states when $u_{2L-SVM} \approx -0.5$. Therefore, the phase shift of the current has no influence on the loss distribution. Also, by alternating the reference signal u_{2L-SVM} between the upper carrier band and lower carrier band, a better loss distribution is achieved for a small modulation index.

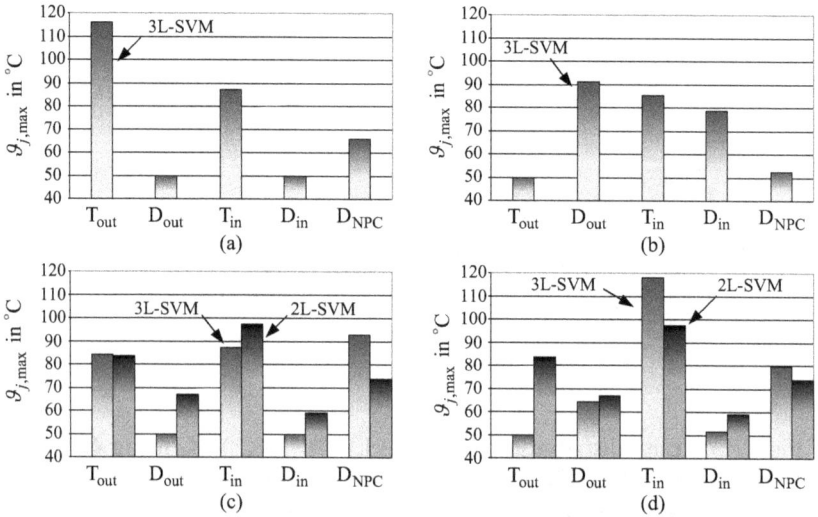

Figure 3.10 Maximal junction temperature distribution in the 3.3 kV 3L-NPC VSC using 4.5 kV PP Westcode T1200EB45E IGBTs and 4.5 kV PP Infineon D1031SH45T diodes; U_{dc}=5000 V, $i_{ph,rms}$=1200 A, f_{sw}=450 Hz, f_1=50 Hz, $\vartheta_a = 50°C$; (a) pf=1, m_a=1.15; (b) pf=-1, m_a=1.15; (c) pf=1, m_a=0.05; (d) pf=-1, m_a=0.05

42

Figure 3.10 shows the distribution of the maximal junction temperatures in the 3L-NPC VSC. The maximal junction temperature describes the maximum value of the junction temperature of a power semiconductor during one period of the fundamental output frequency (see Section 4.2.2). The PP IGBT and the PP diode are installed on separate heat sinks which results in a good thermal decoupling of the semiconductors. Hence, the temperature distribution is similar to the loss distribution. It is assumed that the junction temperature of the semiconductors without losses equals the ambient temperature.

Table 3.7 summarizes the most stressed devices for each critical operation point. The active switches T_{out} in Case A and T_{in} Case D (3L-SVM) present an increased maximal junction temperature, compared to the rest of the power devices. Applying the 2L-SVM in case D, the maximal junction temperature of T_{in} is advantageously reduced.

Table 3.7 Most thermal stressed devices of the 3L-NPC VSC

	Modulation scheme	Most stressed devices	Maximal junction temperature in °C
Case A	3L-SVM	T_{out}	115.9
Case B	3L-SVM	D_{out}	91.2
Case C	3L-SVM	D_{NPC}	92.8
Case D	3L-SVM	T_{in}	118.2
Case C/D	2L-SVM	T_{in}	97.5

3.2 The 3L-ANPC VSC

3.2.1 Structure of the 3L-ANPC VSC

The topology of the 3L-ANPC VSC is similar to the topology of the 3L-NPC VSC with active switches placed in antiparallel to the NPC diodes. The NPC active switches are referred as T_{NPC}. They have the same current and voltage rating as the other switches T_{out} and T_{in} [9]. Figure 3.11 presents the topology of the 3L-ANPC VSC.

43

3. Three-Level Neutral Point Clamped Voltage Source Converter

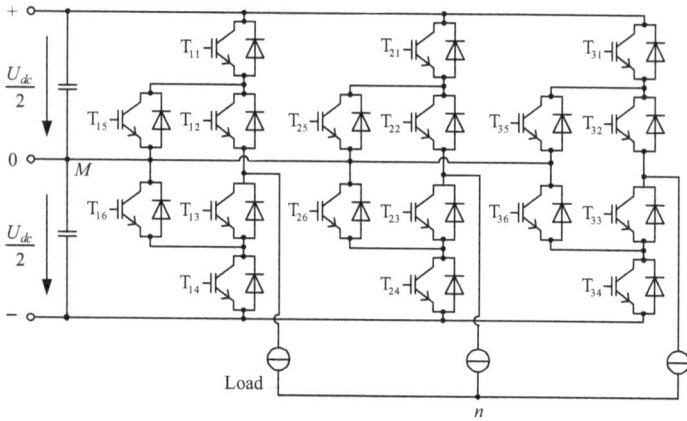

Figure 3.11: Three-Level Active Neutral Point Clamped Voltage Source Converter [9]

As previously reported in [71], the active NPC devices were proposed to guarantee an equal static voltage sharing between the power devices T_{x1} and T_{x2} and between T_{x3} and T_{x4}, respectively. In the 3L-NPC VSC, the equal static voltage sharing is ensured by a balancing resistor across the inner switches. Thus, the employment of the NPC active switches would avoid the balancing resistor and the losses generated in it [9]. The active NPC switches are also applied in case of series connected IGBTs, in order to ensure the dynamic voltage sharing during the turn-off of the series connected NPC diodes [72]. In commercially available MV drives with IGBT modules, the NPC diodes were replaced by IGBT modules with the aim of achieving modularization and standardization. If no active clamping is required, the active switch within the IGBT module is turned off by an external Gate-Emitter-short circuit and thus, the parallel diode acts as the NPC diode [9].

In contrast to the 3L-NPC VSC, new switch states and commutations are obtained by using the active NPC switches. The new available switch states and commutations enable a proper utilization of the upper and lower path of the neutral tap, in order to advantageously distribute losses in the converter.

44

3. Three-Level Neutral Point Clamped Voltage Source Converter

3.2.2 3L-ANPC VSC switch states and commutations

3.2.2.1 Switch states

During the "+" state, T_{x1} and T_{x2} are turned on, while T_{x3} and T_{x4} are turned off. Additionally, T_{x6} should be turned on to ensure an equal voltage sharing between T_{x3} and T_{x4}, while T_{x5} must be in the off-state. Similarly, the "−" state is accomplished by turning on the lower two switches T_{x3} and T_{x4} and also T_{x5}, to guarantee an equal voltage sharing between T_{x1} and T_{x2}. Compared to the 3L-NPC VSC, the 3L-ANPC VSC features four available zero switch states that are referred as "0U2", "0U1", "0L2" and "0L1".

Table 3.8 summarizes the definition of the available switch states. "1" designates on-state and "0" designates off-state of the switch.

Table 3.8 Definition of switch states for the 3L-ANPC VSC [9]

	T_{x1}	T_{x2}	T_{x3}	T_{x4}	T_{x5}	T_{x6}
+	1	1	0	0	0	1
0U2	0	1	0	0	1	0
0U1	0	1	0	1	1	0
0L1	1	0	1	0	0	1
0L2	0	0	1	0	0	1
−	0	0	1	1	1	0

Figure 3.12 shows the conduction paths of the 3L-ANPC VSC. The "0U2" state is achieved when the devices T_{x2} and T_{x5} are turned on, allowing the current to flow through the upper path in both directions. The switch T_{x4} can be turned on or off. By turning on T_{x4}, a new state named "0U1" is achieved. Although T_{x4} is turned on, the conduction path does not change. However, the voltage distribution is changed. Similarly, during the "0L2" state, T_{x3} and T_{x6} are turned on. T_{x1} may be in the on- or off-state. The new switching state "0L2" is achieved by turning on T_{x1}.

45

Another possibility to accomplish the zero state is to turn-on simultaneously T_{x2}, T_{x3}, T_{x5} and T_{x6}. In this case, it is not sure that the current is equally distributed between the upper and lower NPC path. The current distribution is determined by different factors like: the on-state characteristics of the devices, the prior switch state, the junction temperature, and the parasitic stray inductances [9]. Therefore, this state is not further considered.

The distribution of conduction losses between the inner and NPC devices during the zero states can be controlled by the selection of the upper or the lower NPC path. However, the distribution of conduction losses during the "+" and "–" states cannot be influenced.

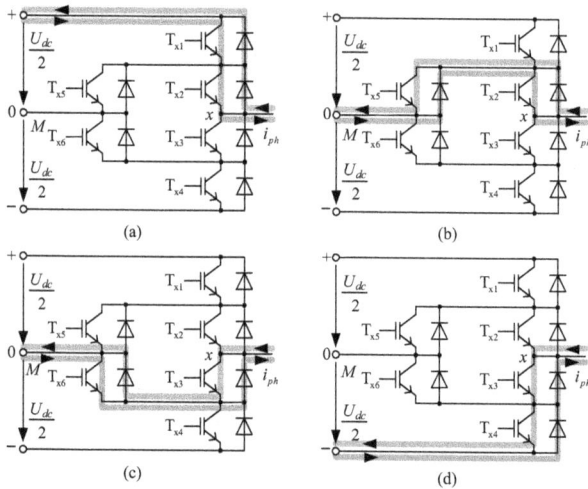

Figure 3.12: Conduction paths of the 3L-ANPC VSC [9]: (a) State "+"; (b) State "0U2", "0U1"; (c) State "0L2", "0L1"; (d) State "–"

Table 3.9 presents the conduction losses of the 3L-ANPC VSC in one phase leg. Similar to the 3L-NPC VSC, there are always two power semiconductors conducting the phase current: two active switches or two diodes in the "+" and "–" states, depending on the direction of the phase current, and one active switch and one diode in

46

the zero state. The total amount of conduction losses is the same for all zero states; only the distribution of losses is different.

Table 3.9 Conduction losses in the 3L-ANPC VSC [9]

	T_{x1}	D_{x1}	T_{x2}	D_{x2}	T_{x3}	D_{x3}	T_{x4}	D_{x4}	T_{x5}	D_{x5}	T_{x6}	D_{x6}
Positive phase current												
+	×		×									
0U2			×							×		
0U1			×							×		
0L1					×					×		
0L2					×					×		
−					×		×					
Negative phase current												
+	×		×									
0U2			×					×				
0U1			×					×				
0L1					×							×
0L2					×							×
−					×		×					

3.2.2.2 Commutations types

The 3L-ANPC VSC presents three types of commutations, depending on the switching loss distribution in one phase leg:

- Type 1 commutation: outer devices with NPC devices;
- Type 2 commutation: outer devices with inner devices;
- Type 3 commutation: inner devices with inner devices.

The Type 1 commutation loop is formed between the positive dc-rail and the upper NPC path and between the negative dc-rail and the lower NPC path. Both Type 2 and Type 3 commutations are formed between one dc-rail with the farther path of the neutral tap, i.e. the positive dc-rail with the lower path or the negative dc-rail with the upper path.

The distribution of the switching losses can be controlled by selecting the commutation from "+" and "−" states to the zero states and backwards. The selection of different zero states has no effect on the

47

modulation of the 3L-ANPC VSC, i.e. voltage waveforms and neutral-point potential control [9]. Figure 3.13 illustrates the commutations for a positive phase current.

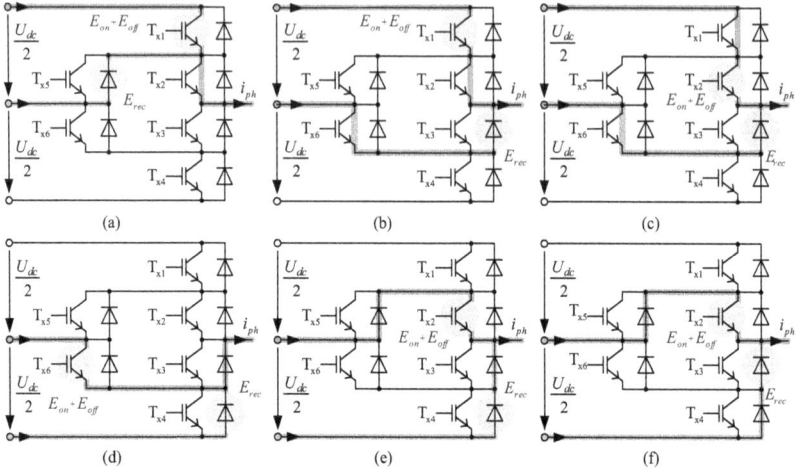

Figure 3.13 Commutations for positive phase current [9]:
(a) "+"↔"0U2"/"0U1"; (b) "+"↔"0L2"; (c) "+ ↔"0L1";
(d) "−"↔"0L2"/"0L1"; (e) "−"↔"0U2"; (f) "−"↔"0U1";

During the **commutation "+"→"0U2"** (see Figure 3.13 (a)), the positive phase current commutates to the upper path of the neutral tap. The switching transitions take place in the following order: T_{x6} has to be turned off first, then T_{x1} is turned off and after a dead time, T_{x5} is turned on. The positive phase current commutates to the NPC diode D_{x5} after T_{x1} is turned off. T_{x1} experiences turn-off losses. During the inverse **commutation "0U2"→"+"**, T_{x5} is turned off first, then T_{x1} is turned on and finally, T_{x6} is turned on. As soon as T_{x1} is turned on, the current commutates back to the positive dc-rail. T_{x1} experiences turn-on losses and D_{x5} recovery losses. Since the switching losses occur in the outer device and the NPC device, the commutation "+"↔"0U2" is considered as Type 1 commutation.

The **commutation "+"→"0U1"** (see Figure 3.13 (a)) is similar to "+"→"0U2", only that T_{x4} is turned on after T_{x5} is turned on. The

48

device T_{x4} does not experience any turn-on losses. The turn-on of T_{x4} does not significantly affect the loss distribution. Therefore, the commutation is not used. However, the reverse **commutation** **"0U1"→"+"** is used in the transition of the modulation of a negative voltage (similar to "0L1" used for the modulation of a positive voltage). The switching transitions take place in the following order: first, T_{x4} has to be turned off. No forward current is flowing through the outer switch T_{x4} and thus, the device does not take any blocking voltage. If T_{x1} is turned on, the inner switch T_{x3} has to block the entire U_{dc} voltage, resulting in an over-voltage that can destroy the power semiconductor. In order to avoid this dangerous over-voltage, after T_{x4} is turned off, T_{x6} is turned on. Thus, the phase current flows simultaneously in the upper path (D_{x5} and T_{x2}) and the lower path (T_{x6} and D_{x3}) of the neutral tap. By turning on T_{x6}, T_{x4} blocks $U_{dc}/2$. After T_{x5} is turned off, T_{x1} is turned on, forcing the current to commutate on the positive dc-rail. T_{x1} and D_{x5} experience turn-on and recovery losses, respectively. T_{x6} and D_{x3} experience insignificant losses. The commutation "+"↔"0U1" is a Type 1 commutation because the switching losses occur in the outer device and the NPC device.

During the **commutation** **"+"→"0L2"** (see Figure 3.13 (b)), the phase current is commutated to the lower path of the neutral tap. The switching transitions take place in two steps. In the first step, T_{x1} is turned off and T_{x3} is turned on after a dead time. In this stage, the positive current commutates to both upper and lower path of the neutral tap because T_{x6} is in on-state. The current distribution is determined by the stray inductances of the commutation paths. The upper path takes the larger part of the current because of the lower stray inductance provided through D_{x5}. The outer switch T_{x1} experiences turn-off losses. In the second step, T_{x2} is turned off after a dead time forcing the positive current to commutate to the lower path of the neutral path. Since T_{x2} is turned off at almost zero voltage, the switching losses are insignificant. In the case of reverse **commutation** **"0L2"→"+"**, the switching transitions take place in the following order: T_{x2} is turned on first and then T_{x3} is turned off. Finally, T_{x1} is turned on. After this turn-on transient, the positive current commutates back to the positive dc-rail. T_{x1} and D_{x3} experience turn-on and recovery losses, respectively. The

49

switching losses occur in the outer device and inner device and thus, the commutation "+"↔"0L2" is a Type 2 commutation.

With the **commutation "+"→"0L1"** (see Figure 3.13 (c)), the phase current is commutated to the lower path of the neutral tap, similar to the commutation "+"→"0L2". In contrast to the commutation "+"→"0L2", T_{x1} remains in the on-state. Only T_{x2} is turned off and then T_{x3} is turned on. The switch T_{x2} experiences turn-off losses. The inverse **commutation "0L1"→"+"** is initiated by the turn-off of T_{x3}. After a dead time, T_{x2} is turned on. Thus, the current commutates back to the positive dc-rail. Switching losses occur in T_{x2} and D_{x3}. Since the switching losses occur in different inner devices, the commutation "+"↔"0L1" is a Type 3 commutation.

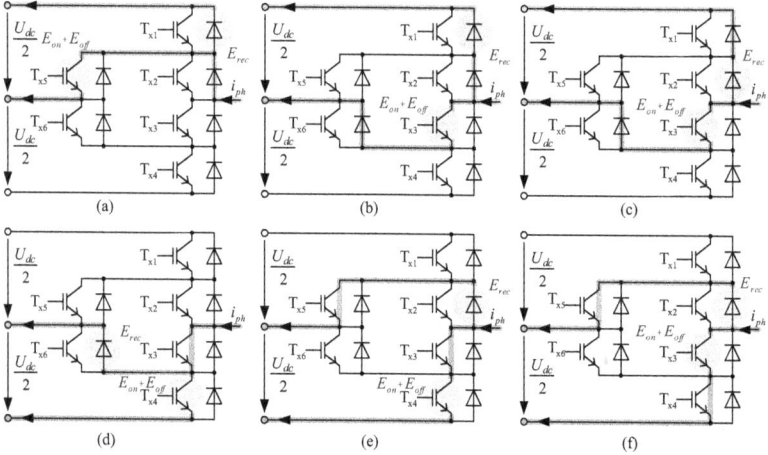

Figure 3.14 Commutations for negative phase current [9]:
(a) "+"↔"0U2"/"0U1"; (b) "+"↔"0L2"; (c) "+ ↔"0L1";
(d) "–"↔"0L2"/"0L1"; (e) "–"↔"0U2"; (f) "–"↔"0U1";

Figure 3.14 illustrates the commutations for a negative phase current. The control signals of the active switches are identical for the commutation for a positive and negative phase current. With the **commutation "+"→"0U2"** (see Figure 3.14 (a)) the negative phase current is commutated to the upper path of the neutral tap. First T_{x6} is

50

turned off, then T_{x1} is turned off and after a dead time, T_{x5} is turned on. T_{x5} and D_{x1} experience turn-on and recovery losses, respectively. During the reverse **commutation "0U2"→"+"**, T_{x5} is turned off first and after a dead time T_{x1} and T_{x6} are turned on. T_{x5} experiences turn-off losses.

As described above, the **commutation "+"→"0U1"** (see Figure 3.14 (a)) differs from the commutation "+"→"0U2" only by the additional lossless turn-on of T_{x4}. Commutations "+"↔"0U2"/"0U1" are considered as Type 1 commutations. During the **commutation "+"→"0L2"** (see Figure 3.14 (b)), the negative phase current is commutated to the lower path of the neutral path. The switching transitions take place in the following order: first, T_{x1} is turned off and after a dead time, T_{x3} is turned on. The NPC switch T_{x6} remains in the on-state. Finally, T_{x2} is turned off to ensure that the negative phase current stays in the lower neutral path in case its direction reverses. T_{x3} experiences turn-on losses, whereas D_{x1} experiences recovery losses. The inverse **commutation "0L2"→"+"** is initiated by the turn-on of T_{x2}, followed by the turn-off of T_{x3}. Thus, the negative current is forced back to the positive dc-rail. Finally, T_{x1} is turned on. The switch T_{x3} experiences turn-off losses. The switching losses occur in the outer device and inner device and thus, the commutation "+"↔"0L2" is considered as Type 2 commutation.

During the **commutation "+"→"0L1"** (see Figure 3.14 (c)), the negative phase current commutates to the lower path of the neutral tap. During this commutation, only T_{x2} is turned off and T_{x3} is turned on after a dead time. T_{x3} and D_{x2} experience turn-on and recovery losses, respectively. The inverse **commutation "0L1"→"+"** is initiated by the turn-off of T_{x3}, forcing the negative current back to the positive dc-rail, followed by the turn-on of T_{x2}. Essential turn-off losses occur in T_{x3}. The commutation "+"↔"0L1" is a Type 3 commutation.

The remaining commutations in Figures 3.13 (d), (e), and (f) and Figures 3.14 (d), (e), and (f) are equivalent to the afore described commutations due to the symmetry of the converter structure.

51

Table 3.10 presents the distribution of the switching losses for all commutations for positive and negative phase current and a summary of the 3L-ANPC commutations types.

Table 3.10 Device switching losses in the 3L-ANPC VSC [9]

	Commutation Type	T_{x1}	D_{x1}	T_{x2}	D_{x2}	T_{x3}	D_{x3}	T_{x4}	D_{x4}	T_{x5}	D_{x5}	T_{x6}	D_{x6}	Remarks
	Positive phase current													
1	+ → 0U2	×												
1	+ ← 0U2	×								×				
1	+ → 0U1	×												Not used
1	+ ← 0U1	×								×				
3	+ → 0L1			×										
3	+ ← 0L1			×		×								
2	+ → 0L2	×												Two-step
2	+ ← 0L2	×				×								
2	0U2 → –			×										
2	0U2 ← –			×					×					
3	0U1 → –			×										
3	0U1 ← –			×		×								
1	0L1 → –											×		
1	0L1 ← –								×			×		Not used
1	0L2 → –											×		
1	0L2 ← –							×				×		
	Negative phase current													
1	+ → 0U2	×						×						
1	+ ← 0U2							×						
1	+ → 0U1	×						×						Not used
1	+ ← 0U1							×						
3	+ → 0L1					×	×							
3	+ ← 0L1					×								
2	+ → 0L2	×				×								
2	+ ← 0L2					×								
2	0U2 → –					×		×						
2	0U2 ← –							×						Two-step
3	0U1 → –					×	×							
3	0U1 ← –					×								
1	0L1 → –							×				×		
1	0L1 ← –							×						Not used
1	0L2 → –					×						×		
1	0L2 ← –					×								

3.2.3 Active Loss Balancing Method

The main goal of the Active Loss Balancing (ALB) method is to equally distribute the converter losses and junction temperatures of the power devices within one phase leg. By a proper selection of the zero states and commutations, the method reduces as much as possible the peak junction temperature of the hottest devices.

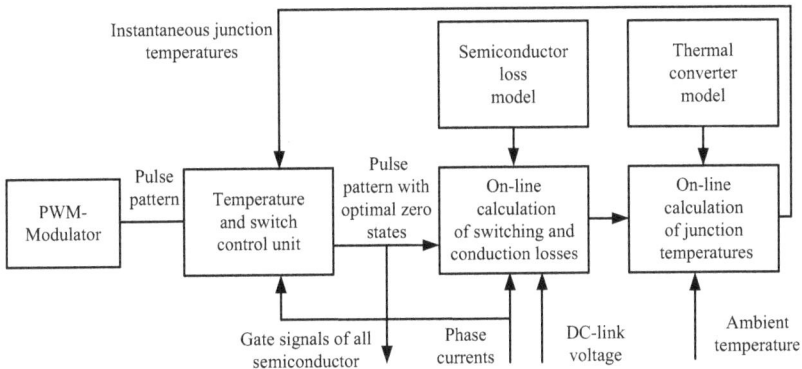

Figure 3.15 Block diagram of the Active Loss Balancing method [9]

The Figure 3.15 shows the block diagram of the Active Loss Balancing (ALB) method. The ALB algorithm is inserted after the PWM modulator for the 3L-NPC VSC. The algorithm estimates on-line the power semiconductors losses and junction temperatures using a semiconductor loss approximation and a thermal model. Depending on the junction temperatures, the decision block named "Temperature and switch control unit" selects the optimal zero state and commutation for the upcoming switching period. The selection of the optimal zero state takes place according to the decision chart presented in Table 3.11, which was proposed in [9]. The "Temperature and switch control unit" generates the gate signals for all 18 power devices of the 3L-ANPC VSC.

53

The optimal zero state is selected during the "+" or "–" states directly before the upcoming commutation. The same type of commutation is then used for the inverse transition back to the same dc-rail. The selection of the zero states and commutation has no effect on the modulation of the 3L-ANPC VSC, i.e. voltage waveforms and neutral-point potential control. Also, the converter total losses remain unchanged [9].

Table 3.11 Decision chart of the Active Loss Balancing method [9]

Modulation	Phase current	Junction temperatures			Zero state
Positive voltage $+\rightarrow 0$ $u_{ref} > 0$	$i_{ph} > 0$	$\vartheta_{jTx1} > \vartheta_{jTx2}$		$\vartheta_{jDx5} > \vartheta_{jDx3}$	OL1
			$\vartheta_{jDx5} < \vartheta_{jDx3}$	$\vartheta_{jTx1} > \vartheta_{jDx3}$	OL1
				$\vartheta_{jTx1} < \vartheta_{jDx3}$	0U2
		$\vartheta_{jTx1} < \vartheta_{jTx2}$		$\vartheta_{jDx5} > \vartheta_{jDx3}$	OL2
				$\vartheta_{jDx5} < \vartheta_{jDx3}$	0U2
	$i_{ph} < 0$	$\vartheta_{jDx1} > \vartheta_{jDx2}$		$\vartheta_{jTx5} > \vartheta_{jTx3}$	OL1
			$\vartheta_{jTx5} < \vartheta_{jTx3}$	$\vartheta_{jDx1} > \vartheta_{jTx3}$	OL1
				$\vartheta_{jDx1} < \vartheta_{jTx3}$	0U2
		$\vartheta_{jDx1} < \vartheta_{jDx2}$		$\vartheta_{jTx5} > \vartheta_{jTx3}$	OL2
				$\vartheta_{jTx5} < \vartheta_{jTx3}$	0U2
Negative Voltage $0\rightarrow -$ $u_{ref} < 0$	$i_{ph} > 0$	$\vartheta_{jTx2} > \vartheta_{jTx6}$	$\vartheta_{jDx4} > \vartheta_{jDx3}$	$\vartheta_{jTx2} > \vartheta_{jDx4}$	OL2
				$\vartheta_{jTx2} < \vartheta_{jDx4}$	0U1
			$\vartheta_{jDx4} < \vartheta_{jDx3}$		OL2
		$\vartheta_{jTx2} < \vartheta_{jTx6}$		$\vartheta_{jDx4} > \vartheta_{jDx3}$	0U1
				$\vartheta_{jDx4} < \vartheta_{jDx3}$	0U2
	$i_{ph} < 0$	$\vartheta_{jDx2} > \vartheta_{jDx6}$	$\vartheta_{jTx4} > \vartheta_{jTx3}$	$\vartheta_{jDx2} > \vartheta_{jTx4}$	OL2
				$\vartheta_{jDx2} < \vartheta_{jTx4}$	0U1
			$\vartheta_{jTx4} < \vartheta_{jTx3}$		OL2
		$\vartheta_{jDx2} < \vartheta_{jDx6}$		$\vartheta_{jTx4} > \vartheta_{jTx3}$	0U1
				$\vartheta_{jTx4} < \vartheta_{jTx3}$	0U2

Table 3.11 presents an algorithm that selects the optimal zero state. According to Table 3.11, for a positive voltage and a positive phase current, the switching losses can be shifted between T_{x1} and T_{x2} and between D_{x3} and D_{x5}.

If the condition $\vartheta_{jTx1} > \vartheta_{jTx2}$ and $\vartheta_{jDx5} > \vartheta_{jDx3}$ is fulfilled, then the optimal zero state is "0L1". By selecting "0L1", T_{x1} and D_{x5} are released from the upcoming switching losses which are now transferred to T_{x2} and D_{x3}, respectively.If the instantaneous junction temperatures satisfy the condition $\vartheta_{jTx1} > \vartheta_{jTx2}$ and $\vartheta_{jDx5} < \vartheta_{jDx3}$, there are two possible zero states "0L1" and "0U2". The selection of "0L1" releases T_{x1} from the upcoming switching losses, which are transferred to T_{x2}. However, this selection transfers the recovery losses from D_{x5} to D_{x3}, which presents a higher junction temperature ($\vartheta_{jDx5} < \vartheta_{jDx3}$). The selection of "0U2" releases D_{x3} from the upcoming recovery losses, which are transferred to D_{x5}. However, this selection transfers the switching losses from T_{x2} to T_{x1}, which presents a higher junction temperature ($\vartheta_{jTx1} > \vartheta_{jTx2}$). To eliminate this ambiguity, the temperature of T_{x1} and D_{x3} are compared. If $\vartheta_{jTx1} > \vartheta_{jDx3}$, than T_{x1} is the most stressed device and the optimal zero state is "0L1". In the other case, if $\vartheta_{jTx1} < \vartheta_{jDx3}$, the inner diode D_{x3} is the hottest power device and the optimal zero state is "0U2".

In the last two cases, if the instantaneous junction temperatures satisfy the condition $\vartheta_{jTx1} < \vartheta_{jTx2}$ and $\vartheta_{jDx5} > \vartheta_{jDx3}$, "0L2" is selected. Thus, the upcoming switching losses are transferred from T_{x2} to T_{x1}, whereas the recovery losses are transferred from D_{x5} to D_{x3}. In the other case, when $\vartheta_{jDx5} < \vartheta_{jDx3}$, the optimal zero state is "0U2".

The loss and junction temperature distribution in the 3.3 kV 3L-ANPC VSC using 4.5kV PP IGBTs and 4.5kV PP diodes is analyzed applying the ALB algorithm. The simulation parameters are summarized in Table 3.6 (Section 3.1.4).

55

Figure 3.16 presents the comparison of loss distributions in the 3L-ANPC VSC for the operations with and without the ALB algorithm. Figure 3.16 (a) presents case A ($pf=1$, $m_a=1.15$). In this case, the ALB algorithm distributes almost equally the switching losses of T_{out} between T_{out} and T_{in}. In this way, T_{out} is released from a significant amount of switching losses. Also, the diode reverse recovery losses are equally allocated between D_{NPC} and D_{in}. The distribution of switching losses is achieved by a combination of Type 1 and Type 3 commutations. The use of Type 2 commutation does not affect the balance between T_{out} and T_{in}, however it increases the imbalance between D_{NPC} and D_{in}. Since the goal of the ALB method is thermal balancing, the Type 2 commutation is not required.

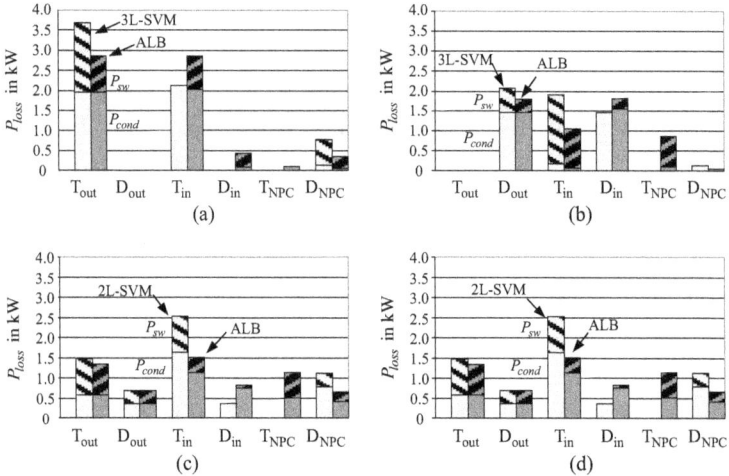

Figure 3.16 Loss distribution in the 3.3 kV 3L-ANPC VSC using 4.5kV PP Westcode T1200EB45E IGBTs and 4.5kV PP Infineon D1031SH45T diodes, $U_{dc}=5000$ V, $i_{ph,rms}=1200$ A, $f_{sw}=450$ Hz, $f_1=50$ Hz, $\vartheta_a = 50°C$; (a) $pf=1$, $m_a=1.15$; (b) $pf=-1$, $m_a=1.15$; (c) $pf=1$, $m_a=0.05$; (d) $pf=-1$, $m_a=0.05$

Figure 3.16 (b) presents the loss distribution in case B ($pf=-1$, $m_a=1.15$). Only Type 1 and Type 3 commutations are employed to achieve a balance between D_{NPC} and D_{in} and between T_{NPC} and T_{in}.

The principle of operation of the ALB algorithm for small modulation index is analyzed only for 2L-SVM, because this modulation distributes more equally the losses and the junction temperatures compared to the 3L-VSC. As mentioned before, the 2L-SVM allocates the losses similarly for case C (see Figure 3.16 (c)) and case D (see Figure 3.16 (d)). The loss balancing achieved with 2L-SVM is further improved by the ALB method. In the cases C and D, all three types of commutations are used. The combination of Type 1 and Type 2 commutations ensures a loss distribution between D_{in} and D_{NPC} and between T_{in} and T_{NPC}, whereas the combination of Type 1 and Type 3 commutations allocates the loss between T_{in} and T_{out}.

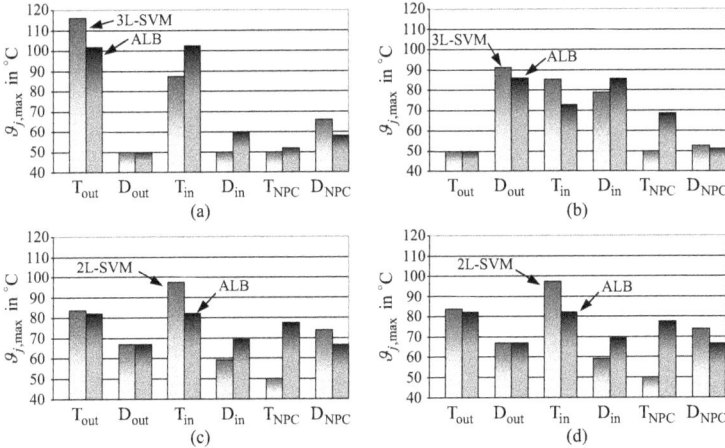

Figure 3.17 Maximal junction temperature in the 3.3 kV 3L-ANPC VSC using 4.5kV PP Westcode T1200EB45E IGBTs and 4.5kV PP Infineon D1031SH45T diodes; U_{dc}=5000 V, $i_{ph,rms}$=1200 A, f_{sw}=450 Hz, f_1=50 Hz, $\vartheta_a = 50°C$; (a) pf=1, m_a=1.15; (b) pf=-1, m_a=1.15; (c) pf=1, m_a=0.05; (d) pf=-1, m_a=0.05

Figure 3.17 illustrates the comparison of the maximal junction temperature distribution in the 3L-ANPC VSC for the operation with and without ALB algorithm. Figure 3.17 (a) illustrates case A, where the ALB method reduces the maximal junction temperature of T_{out} from 115.9 °C to 101.7 °C. The thermal balance between T_{out} and T_{in} and

57

between D_{NPC} and D_{in} is accomplished through the combination of the Type 1 and Type 3 commutations.

Figure 3.17 (b) illustrates case B where the ALB method realizes a loss distribution between D_{out} and D_{in} and between T_{in} and T_{NPC}. The maximal junction temperature of D_{out} is reduced from 91.2 °C to 85.9 °C by using both Type 1 and Type 3 commutations.

Figures 3.17 (c) and (d) illustrate the cases C and D. The maximal junction temperature of T_{in} is reduced from 97.5 °C to 81.9 °C if the ALB method is applied. All three types of commutations, Type 1, Type 2, and Type 3 are required to reduce the maximal junction temperature in the converter at different operating conditions.

3.3 Summary

In this chapter the 3L-NPC VSC and the 3L-ANPC VSC have been presented. The additional switch states and commutations within the 3L-ANPC provide an additional degree of freedom. This gives the possibility to select the optimal zero state with the aim to improve the loss balancing within the converter. The Active Loss Balancing (ALB) algorithm [9] uses a feedback control system, where the controlled variables are the junction temperatures. The algorithm is based on a decision chart. It considers only the switching losses to select the optimal commutation and zero state. In this way, the switching losses of the hottest device are transferred to an alternative switching device until equilibrium between their junction temperatures is achieved. Since the junction temperature limits the maximal allowable output power of the converter, a more symmetrical distribution of the junction temperature offers the possibility to increase the output power range.

After analyzing the ALB method, the question arises if there is a way to obtain a better loss balancing method to reduce as much as possible the highest junction temperature. Can an improved method be developed if not only the switching losses are taking into account, but also the conduction losses and the thermal behavior of the devices? In

58

order to investigate this concept, a new prediction algorithm is proposed. The main idea is to pre-calculate the junction temperature for all possible commutations and zero states for a certain time interval. Similar to the ALB method, the new proposed algorithm searches the optimal zero state. The difference is that the prediction algorithm considers in the selection of the zero state both the conduction and the switching losses and not only the switching losses. Another difference is that the predictive algorithm uses a cost function to evaluate and to determine the optimal zero state. An advantage of the predictive algorithm is that the cost function can be adapted to different requirements. Finally, the cost function with the most promising results is implemented and evaluated in comparison with the ALB method and the conventional NPC converter.

4 Calculations of Power Semiconductor Losses and Junction Temperatures

The first part of this chapter describes a general method to calculate the semiconductor switching and conduction losses. The second part briefly characterizes the thermal model of a power semiconductor and presents a general method to calculate junction temperatures. Finally, the thermal behavior of the 3L-NPC VSC is analyzed at different operation points.

4.1 Calculations of Switching and Conduction Losses

Power dissipation is generated during the dynamic and static state of the power device. The total power loss of a semiconductor can be expressed as [73]:

$$P_{tot} = \underbrace{P_{cond} + P_{bl}}_{\text{Static losses}} + \underbrace{P_{on} + P_{off}}_{\text{Switching losses}} + \underbrace{P_{drive}}_{\text{Driving losses}}, \tag{4.1}$$

where P_{cond} are the conduction losses, P_{on} are the turn-on losses, and P_{off} are the turn-off losses. The blocking losses P_{bl} and the driving losses P_{drive} may usually be neglected since they are small compared with the other types of losses [73].

The power losses of the IGBT and the diode can be expressed as [73], [74]:

$$P_{loss,T} = \underbrace{\frac{t_1}{T_{sw}} \cdot U_{CE,n} \cdot I_{C,n}}_{\text{conduction losses}} + \underbrace{f_{sw} \cdot (E_{on} + E_{off})}_{\text{switching losses}}, \tag{4.2}$$

$$P_{loss,D} = \underbrace{\left(1 - \frac{t_1}{T_{sw}}\right) \cdot U_{F,n} \cdot I_{F,n}}_{\text{conduction losses}} + \underbrace{f_{sw} \cdot E_{rec}}_{\text{switching losses}}, \tag{4.3}$$

4. Calculations of Power Semiconductor Losses and Junction Temperatures

where $d_T = \frac{t_1}{T_{sw}}$ is the duty cycle of the IGBT and $d_D = 1 - \frac{t_1}{T_{sw}}$ is the duty cycle of the diode. T_{sw} denotes the period of the switching frequency and t_1 denotes the on-time of the IGBT. The IGBT and diode conduction losses are calculated using the on-state saturation voltage $U_{CE,n}$ and $U_{F,n}$, respectively, and the collector current $I_{C,n}$ and the forward current $I_{F,n}$, respectively. The switching losses depend on the switching frequency f_{sw} and the switching energies: E_{on} (turn-on energy), E_{off} (turn-off energy), and E_{rec} (recovery energy).

The semiconductors losses can be approximated by analytical expressions based on the information extracted from data sheets or from measurements (see Section 5.2).

4.2 Thermal Networks and Junction Temperatures

The most important thermal aspects in the design of a power converter are the maximal junction temperature and the junction temperature fluctuation [75]. The maximal junction temperature specified in data sheets for ensuring the safe operation of the device is usually between 125 °C and 175 °C, depending on the device type. Exceeding this temperature might not result in an immediate failure; however, it effects negatively the reliability and the lifetime of the power devices [76]. Also, the device manufactures will not guarantee the parameters specified in the data sheets if the maximal junction temperature is exceeded.

Not only the maximal allowable junction temperature influences the lifetime of the semiconductor, but also the junction temperature fluctuation [77], [78]. The rise and fall of temperature leads to thermal alternating stress causing a degradation of the semiconductor. Failures like lifting of bond wires and solder joint fatigue are caused by the thermo-mechanical stress imposed by different coefficients of thermal expansion (CTE) of the used materials within power modules (see Figure 4.1). Basically there are two types of test to determine the load cycle capability: thermal cycling and power cycling [77]. The thermal cycling

61

4. Calculations of Power Semiconductor Losses and Junction Temperatures

test is performed by externally heating and cooling the module. During thermal cycles the excursions of the case temperature are measured with the aim to test the thermal fatigue of the solder structure. The power cycling test is performed by internally heating, i.e. by powering the module periodically. During power cycling tests the junction temperature is raised and lowered within defined time intervals. The aim of this test is to evaluate the quality of the bond wire and the solder joint [78].

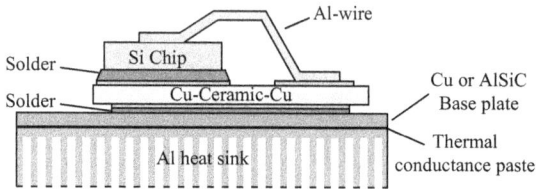

Figure 4.1 The classical layer construction of an IGBT power module [79]

The calculation of the junction temperature depends on the inner structure of the power device, the mounting on a heat sink and the type of the heat sink. The thermal behavior of a power device can be analyzed by network simulation methods [76], [80]. Alternatively, numerical methods like the finite difference method [81] can be employed to determine the temperature.

Table 4.1 Equivalence between thermal and electrical models [82]

Thermal		Electrical	
Temperature	ϑ in K	Voltage	U in V
Heat flow	P in W	Current	I in A
Thermal resistance	R_{th} in K/W	Resistance	R in V/A
Thermal capacitance	C_{th} in Ws/K	Capacitance	C in As/V

The physical mass structure of a power device can store the heat through thermal capacitance C_{th} and can dissipate the heat towards the ambient trough the thermal resistance R_{th}. The heat conduction process can be modeled using the analogy between thermal and electrical processes and parameters [82], [80]. Thus, the thermal resistance and thermal capacitance are equivalent to the electrical resistance and electrical capacitance, respectively. Likewise, the heat flow is analogous to the electrical current, while the temperature is the equivalent to the voltage (see Table 4.1).

The power semiconductor can be subdivided into individual layers and each of them is characterized trough a thermal capacitance C_{th} and a thermal resistance R_{th}. Thus, the resulting equivalent electrical circuit presents several R_{th} and C_{th} elements in order to accurately describe the heat conduction process. The junction-to-heat sink R_{th} and C_{th} elements are influenced by the structure of the chip, the DCB substrate, solder and adhesive, base plate, and the assembly of the module (thermal contact to the heat sink and thermal paste or foil) [73].

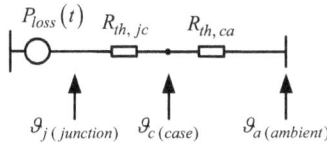

Figure 4.2 Thermal network featuring steady-state resistances [80]

Commonly, the steady-state thermal resistance R_{th} (see Figure 4.2) is utilized to calculate the average junction temperature of a power device [80]:

$$\vartheta_{j,avg} = R_{th,jc} \cdot P_{loss} + R_{th,ca} \cdot P_{loss} + \vartheta_a ,\qquad (4.4)$$

where $R_{th,jc}$ is the thermal resistance junction-to-case, $R_{th,ca}$ is the thermal resistance case-to-ambient, P_{loss} are the total losses, and ϑ_a is the ambient temperature.

This is a practical method to apply at stationary operating points like e.g. a dc current. However, this static model is not suited if the power device is working under pulsed power operation [83]. In this case the dynamic thermal model has to be applied, which is obtained considering heat conduction and storage by the use of R_{th} and C_{th} elements [82].

4.2.1 Foster and Cauer Networks

The thermal behavior of a power module can be characterized using equivalent circuit models. Figure 4.3 [84] illustrates the most used thermal equivalent circuits: Cauer (also called continued fraction circuit) and Foster (also called Pi model or partial fraction circuit).

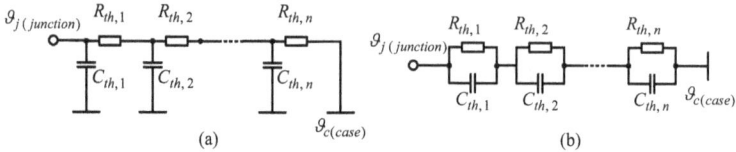

(a)

(b)

Figure 4.3 Equivalent thermal models: (a) Cauer model; (b) Foster model [

The Cauer model [82], [84] describes the real physical setup layer of the power semiconductor. The Cauer model is derived from the transmission line theory and describes the natural heat conduction and storage in a given power device [82]. The model can be obtained from the material characteristics of the layers: chip, chip solder, substrate, substrate solder and base plate. The R_{th} and C_{th} elements represent an individual layer, whereas the network nodes determine the layer sequence. The thermal distribution within the power device can be analyzed using material analysis and finite element method (FEM) simulations [76]. Since the material characteristics are not available in data sheets and are rather difficult to obtain, the design of the Cauer network is laborious.

The Foster model is described by a network of series connected thermal capacitors in parallel with thermal resistances. Because the thermal capacitors are series connected, if a power pulse is applied, the temperature in all internal nodes will change instantaneously. This is clearly different in the physical layers since the heat flow reaches the surface layer with a time delay. Although R_{th} and C_{th} elements of a Foster network have no physical correspondence to an individual layer of the power device, they can be easily extracted from the cooling or heating transient curve specified in data sheets [84], [85].

4. Calculations of Power Semiconductor Losses and Junction Temperatures

The partial fraction coefficients can be found in data sheets in form of thermal resistances R_{th} and the thermal time constants τ_{th}, whereas $\tau_{th} = R_{th} \cdot C_{th}$. The transient thermal impedance is related in system theory to the step response. Thus, the information contained in the thermal impedance curve can be used to evaluate the thermal behavior of the power device for different power inputs. Therefore, the Foster thermal model is easy to evaluate and to implement [86]. The thermal impedance Z_{th} is defined in the time domain for a number n of R_{th} and C_{th} elements as followed:

$$Z_{th}(t) = \sum_{i=1}^{n} R_{th,i} \cdot \left(1 - e^{-\frac{t}{\tau_{th,i}}} \right). \qquad (4.5)$$

The Foster thermal network can be used for thermal model calculations if the junction temperature behavior of the power semiconductor is of main interest and not the exact internal temperature distribution [82].

Issues arise when the thermal model of a power semiconductor has to merge with the thermal model of the heat sink. Usually the power semiconductor and the heat sink are separately characterized from each other, representing two independent thermal systems. Both Foster and Cauer networks can be used to describe the thermal behavior of the power device and the heat sink. The question in this case is: which thermal model approximates better the system composed from semiconductor and heat sink? Since the Cauer model is based on the physical layers, it is allowed to link the thermal systems of the power semiconductor and the heat sink. This method is difficult to implement, because the parameters of the Cauer model need to be determined from the physical structure. An alternative method is to transform the Foster model that is given in data sheets in a Cauer model. There are different methods to mathematically transform the Foster model into Cauer model [87], [88]. The transformation does not give a unique solution and the R_{th} and C_{th} elements depend on how they were extracted from the original characteristics. Also, the obtained Cauer model has no physical significance after the transformation. Therefore, the accuracy of the thermal behavior of the linked Cauer models is negatively affected [84].

Another solution is to merge the Foster models of the power semiconductor and heat sink. For an air-cooled system, the thermal time constants of the heat sink are substantially larger than the values of the power semiconductor. Thus, the calculation of the junction temperature is negatively affected only to a very small degree. Issues arise when water-cooled systems are employed since the thermal model of the heat sink contains comparable small thermal time constants. In this case the thermal impedance Z_{th} of the entire system, power semiconductor and heat sink, has to be measured [84].

4.2.2 Calculation of Junction Temperature

The calculation of the instantaneous junction temperature of a power device is evaluated with the discrete-time state-variables system:

$$\underline{T}(k+1) = \underline{A} \cdot \underline{T}(k) + \underline{B} \cdot \underline{P}_{loss}(k), \qquad (4.6)$$

$$\vartheta_j(k) = \underline{C} \cdot \underline{T}(k) + \vartheta_a, \qquad (4.7)$$

where k is the discrete point at which the system is being evaluated. $\underline{T}$ is the system's vector with elements representing the temperatures across the R_{th} and C_{th} pairs (temperature differences as state variables), $\underline{A}$, $\underline{B}$, and $\underline{C}$ are the system's matrixes. The system matrixes are expressed as:

$$\underline{A} = \begin{pmatrix} 1 - \dfrac{T_{sample}}{C_{th,1} \cdot R_{th,1}} & \cdots & 0 \\ \vdots & \ddots & \vdots \\ 0 & \cdots & 1 - \dfrac{T_{sample}}{C_{th,n} \cdot R_{th,n}} \end{pmatrix}, \quad \underline{B} = \begin{pmatrix} \dfrac{T_{sample}}{C_{th,1}} \\ \vdots \\ \dfrac{T_{sample}}{C_{th,n}} \end{pmatrix}, \quad \underline{C} = (1 \ldots 1) \cdot \qquad (4.8)$$

Figure 4.4 (a) depicts the thermal network of a single power device x (e.g. IGBT or diode) installed on a separate heat sink. In this case, the junction temperature depends on the losses of the power device,

66

4. Calculations of Power Semiconductor Losses and Junction Temperatures

thermal constants of the semiconductor, thermal design of the heat sink, and the ambient temperature. The thermal model of the power semiconductor x is described by a number of n series junction-to-case pair $R_{th,n}$ and $C_{th,n}$. The heat sink thermal model including the thermal grease is described by the case-to-ambient pair $R_{th,ca}$ and $C_{th,ca}$. Figure 4.4 (b) depicts the thermal network of the power transistor T and the parallel inverse diode D installed in the same module. If one semiconductor experiences losses, the antiparallel power device is also heating up because of the thermal coupling via the base plate. An important issue is the mutual heating of the chips due to lateral heat flow in the case of multi-chip modules [89], [76], [80]. Since the dies are in the same module and they are so close to each other, one die can create significant temperature increase for its neighbor. This mutual heating has a major influence on the prediction of junction temperature especially under dc operation mode and at low fundamental frequency [80]. The mutual heating is not considered in the thermal model of Figure 4.4.

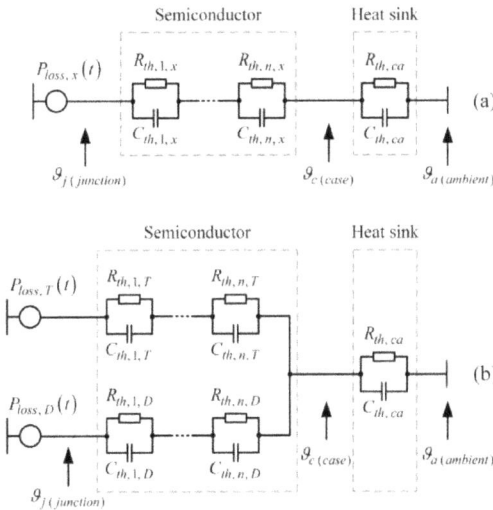

Figure 4.4 Foster thermal network (junction to ambient): (a) single power device installed on a separate heat sink; (b) two power devices installed on a common heat sink

4.2.3 Thermal Design consideration for the 3L-NPC VSC

The junction temperature behavior is an important aspect in the thermal design of MV converters because it affects:
- the switching frequency and the output power of the converter [9];
- the reliability (expected operating lifetime) of the converter due to thermo-mechanical stress [73];
- the mechanical construction of the converter and design of the cooling system, with influence on weight and costs.

Usually, the average junction temperature is used to evaluate the thermal behavior of a power converter in an output frequency range of $f_l \geq 5$ Hz. However, at lower output frequency the junction temperature ripple increases [9], [90]. Thus, the maximal phase current can no longer be determined based on the average junction temperature. Furthermore, large junction temperature ripples affect the reliability of the power semiconductors. To evaluate the thermal performance of the converter, the following critical factors are considered [75]:
- $\vartheta_{j,max}$ (maximal junction temperature);
- $\vartheta_{j,avg}$ (average temperature);
- $\Delta\vartheta_j$ (temperature ripple).

The junction temperature ripple $\Delta\vartheta_j$ (see Figure 4.5) is defined as the difference between the maximal and minimal junction temperature over the fundamental period:

$$\Delta\vartheta_j = \vartheta_{j,max} - \vartheta_{j,min}. \tag{4.9}$$

In the following section, the thermal behavior of a 3.3 kV 3L-NPC VSC is investigated at different operation points (see Table 4.2).

Figure 4.5 shows the thermal behavior of the 3L-NPC VSC at grid frequency (Case 1). Figure 4.5 (a) illustrates the normalized phase-to-NP voltage and the normalized phase current over two fundamental periods. Figure 4.5 (b) shows the thermal behavior of the semiconductors

4. Calculations of Power Semiconductor Losses and Junction Temperatures

conducting during the positive part of the phase current. The semiconductors heat up and cool down over the fundamental period due to the losses, generating the junction temperature ripple $\Delta \vartheta_j$.

Table 4.2 Simulation parameters of a 3.3 kV 3L-NPC VSC using 4.5 kV PP IGBTs

Operating points						Simulation parameters of a 3.3 kV 3L-NPC VSC 4.5 kV PP Westcode T1200EB45E IGBTs and 4.5 kV PP Infineon D1031SH45T diodes		
Case	f_1 in Hz	m_a	PWM scheme	pf	$i_{ph,rms}$ in A	f_{sw} in Hz	U_{dc} in V	ϑ_a in °C	$\vartheta_{j,max}$ in °C
1	50	1.15	3L-SVM	+1	1200				
2	2	0.05	2L-SVM	+1	1200	450	5000	50	125
3	0	0.05	2L-SVM	+1	770				

The amplitude of the junction temperature ripple depends on the thermal constants of the power switches, the output frequency, and the magnitude of the power losses. The jumps in the junction temperature of T_{x1} and D_{x5} are caused by the switching losses (see Table 3.4). The highest temperature ripple occurs in T_{x1}, i.e. 17 K. Values less than 10 K are considered to be uncritical [9].

Another thermal aspect is the difference between the average and the maximal junction temperature. The largest difference occurs in T_{x1}: the maximal junction temperature of $\vartheta_{j,max} = 116$ °C exceeds the average junction temperature $\vartheta_{j,avg} = 105$ °C by 11 K.

Figure 4.6 shows the thermal behavior of the 3L-NPC VSC at low output frequency (Case 2). Figure 4.6 (a) illustrates the reference voltage for 2L-SVM and the normalized phase current. The frequency of the common-mode dc offset is set to $f_{comm} = 25$ Hz. The most thermal stressed semiconductors are T_{x2} and T_{x1}: their junction temperature ripples reach 37 K and 30 K, respectively (see Figure 4.7 (b) and (c)). The junction temperature ripples of D_{x5}, D_{x4} and D_{x3} reach about 14 K (see Figures 4.7

69

4. Calculations of Power Semiconductor Losses and Junction Temperatures

(d), (e) and (f)). Compared to the operation at grid frequency, the temperature ripples are significantly higher. Therefore, to evaluate the performance of the temperature balancing strategy, the junction temperature ripple has to be also considered.

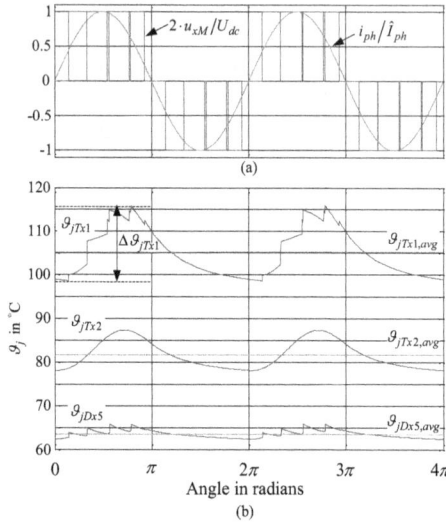

Figure 4.5 Thermal behavior of 3.3 kV 3L-NPC VSC using 4.5 kV PP IGBTs: (a) Normalized phase-midpoint output voltage and normalized phase current, (b) Junction temperatures: ϑ_{jTx1}, ϑ_{jTx2}, ϑ_{jDx5}; U_{dc}=5000 V, $i_{ph,rms}$=1200 A, f_{sw}=450 Hz, f_1=50 Hz, ϑ_a = 50°C ; (a) pf=1, m_a=1.15, 3L-SVM

The difference between the average and the maximal junction temperature significantly increases at low output frequency (e.g. f_1<5 Hz). The largest difference occurs in T_{x1}: the maximal junction temperature of $\vartheta_{j,max}$ = 112 °C exceeds the average junction temperature $\vartheta_{j,avg}$ = 88 °C by 24 K.

70

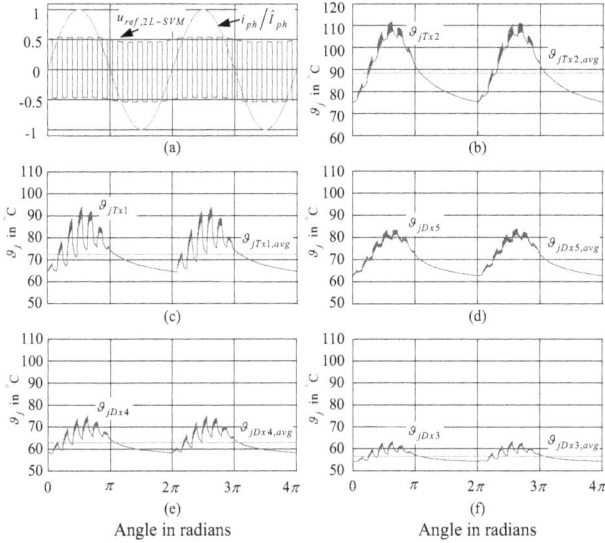

Figure 4.6: Thermal behavior of a 3.3 kV 3L-NPC VSC using 4.5 kV PP IGBTs: (a) corresponding reference waveform for 2L-SVM (f_{comm}=25 Hz), normalized phase current; Junction temperatures: (b) ϑ_{jTx2}, (c) ϑ_{jTx1}, (d) ϑ_{jDx5}, (e) ϑ_{jDx3}, (f) ϑ_{jDx4}; f_1=2 Hz, U_{dc}=5000 V, $i_{ph,rms}$=1200 A, $\vartheta_a = 50°C$, pf=1, m_a=0.05

Figure 4.7 shows in comparison the average and the maximal junction temperature of the power devices of the 3L-NPC VSC. The difference between the average and maximal junction temperature is increasing with decreasing the output frequency. To determine the performance of the temperature balancing strategy, both thermal factors have to be evaluated. Furthermore, because of the considerable temperature difference at lower output frequency, i.e. 24 K, it is not recommended to determine the thermal behavior of the 3L-NPC VSC solely evaluating the average junction temperature.

4. Calculations of Power Semiconductor Losses and Junction Temperatures

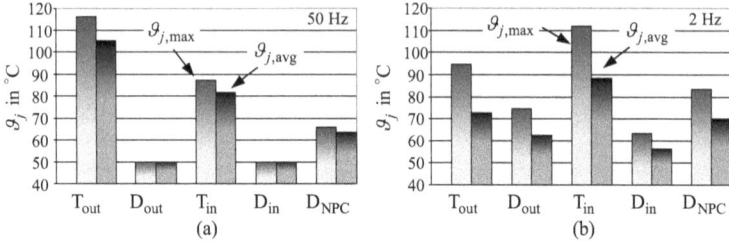

Figure 4.7 Junction temperature distribution of a 3.3 kV 3L-NPC VSC using 4.5 kV PP IGBTs: (a) f_1=50 Hz, pf=1, m_a=1.15, 3L-SVM, **(b)** f_1=2 Hz, pf=1, m_a=0.05, 2L-SVM; U_{dc}=5000 V, $i_{ph,rms}$=1200 A , ϑ_a = 50°C

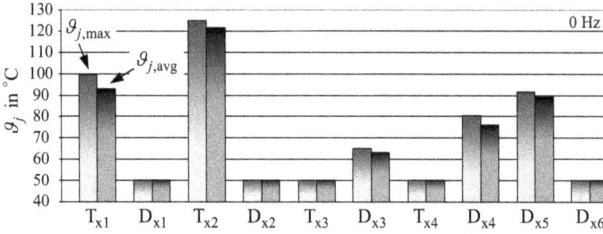

Figure 4.8 Junction temperature distribution in a 3.3 kV 3L-NPC VSC at zero speed: f_1=0 Hz, U_{dc}=5000 V, $i_{ph,rms}$=770 A, ϑ_a = 50°C, pf=1, m_a=0.05, 2L-SVM

Figures 4.8 and 4.9 show the junction temperature distribution of a 3.3 kV 3L-NPC VSC at zero speed (Case 3). The phase current is considered to remain constant at the peak value. The reference voltage for 2L-SVM alternates between the carrier bands with the constant frequency f_{comm}=25 Hz. Thus, for a positive phase current, the power devices T_{x1}, T_{x2}, D_{x3}, D_{x4} and D_{x5} are stressed with losses. In order to keep the maximal junction temperature below 125 °C, the *rms* value of the phase current has to be reduced to 770 A. At zero speed, the temperature ripple is lower than 13 K and the difference between $\vartheta_{j,max}$ and $\vartheta_{j,avg}$ is smaller compared to the aforementioned operating point.

The thermal analysis of the 3L-NPC VSC reveals that not only the average junction temperature has to be considered in the power

72

4. Calculations of Power Semiconductor Losses and Junction Temperatures

converter design, but also the maximal junction temperature and the junction temperature ripple. The decrease of the junction temperature enables an increase of the converter output power and switching frequency. Moreover, reducing the temperature ripple will improve the lifetime expectancy of the power semiconductors.

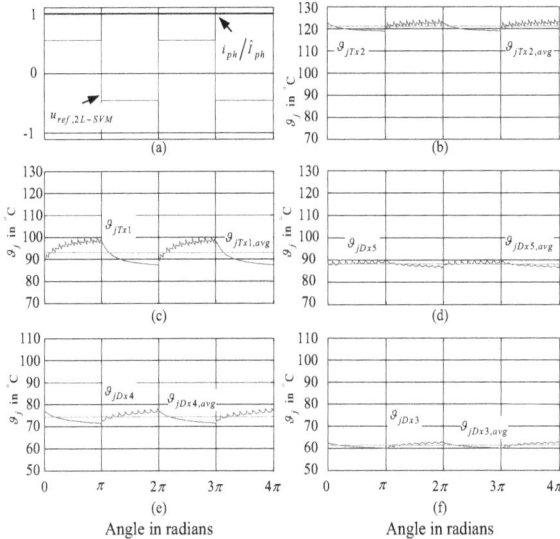

Figure 4.9: Thermal behavior of a 3.3 kV 3L-NPC VSC at zero speed: (a) corresponding reference waveform for 2L-SVM (f_{comm}=25 Hz) and normalized phase current; Junction temperatures: (b) ϑ_{jTx2}, (c) ϑ_{jTx1}, (d) ϑ_{jDx5}, (e) ϑ_{jDx3}, (f) ϑ_{jDx4}; f_1=0 Hz, U_{dc}=5000 V, $i_{ph,rms}$=770 A, $\vartheta_a = 50°C$, pf=1, m_a=0.05

Thus, the Active Loss Balancing method and the Predictive Active Loss Balancing (PALB) method are evaluated in comparison regarding the average and maximal junction temperature, and the junction temperature ripple. The PALB method uses a cost function that can be adapted to different optimization criteria. Thus, the variables for the cost function can be chosen considering one of the three aforementioned thermal factors. Since the most important aim of the PALB method is to reduce the maximal junction temperature, this optimization criteria is chosen as variable for the work in this thesis.

73

5 Predictive Active Loss Balancing (PALB) Method for 3L-ANPC VSCs

A new balancing method is proposed to minimize the highest junction temperature per phase leg in the 3L-ANPC VSC. The goal of the Predictive Active Loss Balancing (PALB) method is to obtain an improved performance of the 3L-ANPC VSC by means of predicting the converter's thermal behavior.

This chapter presents a 3.3 kV 3L-ANPC VSC using 4.5 kV PP IGBTs and PP diodes and a 3.3 kV 3L-ANPC VSC using 4.5 kV IGBT modules. The performance of the converters applying the PALB method is investigated at grid frequency, at low output frequency, and at zero speed. The results are also compared to the standard balancing method, the ALB.

5.1 Description of the PALB method

Compared to the 3L-NPC VSC, the 3L-ANPC VSC presents additional switch states and commutations that enable a proper utilization of the upper and lower path of the neutral tap [9]. Through an optimal selection of commutations and zero states, the loss distribution within the converter can be considerable improved. Moreover, a balanced loss distribution implies a balanced junction temperature distribution, where the maximal junction temperature is advantageously reduced. This achievement offers the potential for an extended output power range and/- or an increased switching frequency, i.e. an improved harmonic performance.

The PALB method selects the optimal zero state during the "+" or "−" states directly before the upcoming commutation, i.e. the commutation "+"→"0" or "−"→"0". The same type of commutation is then used for the inverse transition back to the same dc-rail.

74

Although there are four available zero states "0U2", "0U1", "0L1", and "0L2", the PALB method evaluates only three of them, depending on the sign of the reference voltage. For a positive reference voltage, the commutation "+"→"0U1" differs from the commutation "+"→"0U2" only by the additional lossless turn-on of T_{x4}. Since this does not yield a significant effect on the loss distribution, the zero state "0U1" is not considered. Thus, only commutations from "+" state to "0U2", "0L1", and "0L2" are investigated. Similarly, for a negative voltage reference, the commutation "−"→"0L1" differs from the commutation "−"→"0L2" only by the additional lossless turn-on of T_{x1}. Since there is no influence on the loss distribution, the zero state "0L1" is not considered. Therefore, only commutations from "−" state to "0U2", "0U1", and "0L2" are investigated.

The block diagram of the Predictive Active Loss Balancing (PALB) algorithm is presented in Figure 5.1. The block structure "Converter specification" contains input information about the electrical and thermal parameters of the converter (see Section 5.2). The block "3L-Modulator" contains information about the pulse pattern, the modulation index, and the power factor. Every sample period T_{sample} the module "Loss calculation" computes the instantaneous switching and conduction losses, as shown in Section 4.1. The instantaneous losses are fed into the module "Junction temperatures calculation", that estimates the instantaneous junction temperatures of the 6 IGBTs and the 6 diodes during the current interval k. The decision block "Zero state in $k+1$" verifies if during the next step interval a commutation to the zero state will occur. If there is no commutation to the zero state, the algorithm continues to calculate the losses and the junction temperatures. Otherwise, if a commutation to the zero state will occur during the next step $k+1$, the block "Optimal zero state calculation" selects the optimal zero state x_{op0}. The input information for this block are the junction temperatures at the instant k, the phase current, and the switching pattern until the next zero state.

5. PALB method for 3L-ANPC VSC

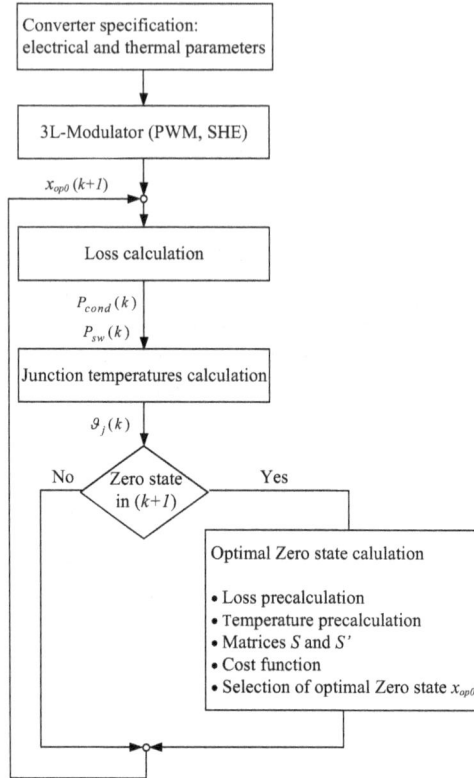

Converter specification:
electrical and thermal parameters

3L-Modulator (PWM, SHE)

$x_{op0}(k+1)$

Loss calculation

$P_{cond}(k)$
$P_{sw}(k)$

Junction temperatures calculation

$\vartheta_j(k)$

No — Zero state in $(k+1)$ — Yes

Optimal Zero state calulation

• Loss precalculation
• Temperature precalculation
• Matrices S and S'
• Cost function
• Selection of optimal Zero state x_{op0}

Figure 5.1 Block diagram of the Predictive Active Loss Balancing (PALB) algorithm

Figure 5.2 shows the graphical representation of the PALB operation principle. The goal of the algorithm is to select the zero state that minimizes the maximal junction temperature during the time interval from $k+1$ until the future commutation to the next zero state, i.e. until $k+n$. This interval represents the prediction time and is designated as $t_{prediction}$ (see hatched area in Figure 5.2). The exemplified pulse pattern is shown during the entire output period T_1.

The algorithm calculates the possible losses during the time interval $t_{prediction}$, for the possible zero state and for each semiconductor. Then, the losses are fed into the thermal model, which estimates the

76

junction temperatures. Afterwards, the algorithm selects the maximal junction temperature over the time interval $t_{prediction}$ for each zero state x and for each semiconductor Sw, as shown in (5.1):

$$\vartheta_{j\_Sw,x} = \max(\vartheta_{j\_Sw,x}(k+1),...,\vartheta_{j\_Sw,x}(k+n)) \cdot \qquad (5.1)$$

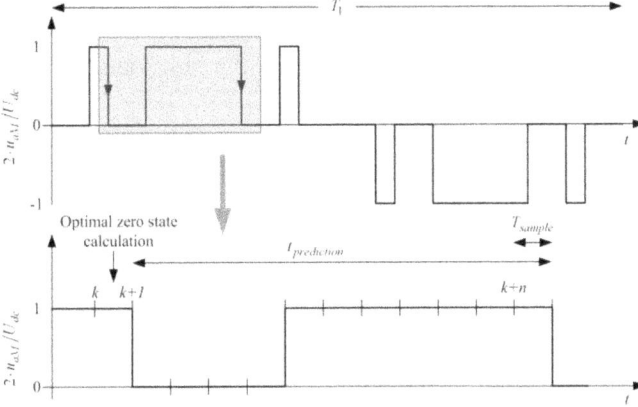

Figure 5.2 Representation of the prediction principle

The maximal junction temperatures are stored in a matrix with 3 rows (3 zero states) and 12 columns (6 IGBTs and 6 diodes), as shown in (5.2). The matrix S is calculated only one time during the entire time interval $t_{prediction}$:

$$S = \begin{bmatrix} \vartheta_{j\_T_1,0U2} & \vartheta_{j\_D_1,0U2} & \cdots & \vartheta_{j\_T_6,0U2} & \vartheta_{j\_D_6,0U2} \\ \vartheta_{j\_T_1,0L2} & \vartheta_{j\_D_1,0L2} & \cdots & \vartheta_{j\_T_6,0L2} & \vartheta_{j\_D_6,0L2} \\ \vartheta_{j\_T_1,0U1/0L1} & \vartheta_{j\_D_1,0U1/0L1} & \cdots & \vartheta_{j\_T_6,0U1/0L1} & \vartheta_{j\_D_6,0U1/0L1} \end{bmatrix} \cdot \qquad (5.2)$$

The algorithm sorts the junction temperatures of the matrix S from the coldest to the hottest device, as shown in (5.3):

$$S' = \begin{bmatrix} \vartheta_{j\_y_1,0U2} & \vartheta_{j\_y_2,0U2} & \cdots & \vartheta_{j\_y_{11},0U2} & \vartheta_{j\_y_{12},0U2} \\ \vartheta_{j\_y_1,0L2} & \vartheta_{j\_y_2,0L2} & \cdots & \vartheta_{j\_y_{11},0L2} & \vartheta_{j\_y_{12},0L2} \\ \vartheta_{j\_y_1,0U1/0L1} & \vartheta_{j\_y_2,0U1/0L1} & \cdots & \vartheta_{j\_y_{11},0U1/0L1} & \vartheta_{j\_y_{12},0U1/0L1} \end{bmatrix}, \qquad (5.3)$$

where $\vartheta_{j\_y_{12},0U2}$ is the maximal junction temperature applying the zero state "0U2", $\vartheta_{j\_y_1,0U2}$ is the minimal junction temperature applying the zero state "0U2", etc.

The evaluation and determination of the optimal zero state is accomplished with a cost function. The cost function goal is to minimize the highest junction temperature during the time interval $t_{predicition}$, as shown in (5.4):

$$g = \min(\vartheta_{j\_y_{12},0U2}, \vartheta_{j\_y_{12},0L2}, \vartheta_{j\_y_{12},0U1/0L1}) \cdot \tag{5.4}$$

An example of how the algorithm selects the optimal zero state x_{op0} is shown in (5.5):

(a) *if* $\vartheta_{j\_y_{12},0U2} < \vartheta_{j\_y_{12},0L2} < \vartheta_{j\_y_{12},0U1/0L1}$

    then $x_{op0} = 0U2$

(b) *if* $\vartheta_{j\_y_{12},0U2} = \vartheta_{j\_y_{12},0L2} < \vartheta_{j\_y_{12},0U1/0L1}$

    then

        $g = \min(\vartheta_{j\_y_{11},0U2}, \vartheta_{j\_y_{11},0L2})$

        if $\vartheta_{j\_y_{11},0U2} < \vartheta_{j\_y_{11},0L2}$

            then $x_{op0} = 0U2$

            else $x_{op0} = 0L2$ (5.5)

        if $\vartheta_{j\_y_{11},0U2} = \vartheta_{j\_y_{11},0L2}$ etc,

        then

            $g = \min(\vartheta_{j\_y_{10},0U2}, \vartheta_{j\_y_{10},0L2})$

(c) *if* $\vartheta_{j\_y_{12},0U2} = \vartheta_{j\_y_{12},0L2} = \vartheta_{j\_y_{12},0U1/0L1}$

    then

        $g = \min(\vartheta_{j\_y_{11},0U2}, \vartheta_{j\_y_{11},0L2}, \vartheta_{j\_y_{11},0U1/0L1})$ etc.,

If the condition **(a)** is satisfied, then the algorithm selects as optimal zero state "0U2" that delivers the lowest junction temperature among the three junction temperatures from the 12th column of matrix S'.

In the same way, the selection algorithm evaluates if "0L2" and "0U1"/"0L1" are the optimal zero state. If the condition **(b)** is satisfied, both "0U2" and "0L2" zero states deliver the same lowest junction temperature among the 12th column of matrix S'. In this case, the cost function g minimizes the junction temperatures among the 11th column of matrix S' delivered by the "0U2" and "0L2" zero states. If this two junction temperatures from the 11th column are equal, then the algorithm searches the lowest junction temperature among the 10th column of matrix S' delivered by the "0U2" and "0L2" zero states and so on. If the condition **(c)** is satisfied, all three zero states deliver the same lowest junction temperature from the 12th column of matrix S'. To select the optimal zero state, the cost function g minimizes the three junction temperatures among the 11th column of matrix S'. This proceeding is repeated until the algorithm finds the optimal zero state. Finally, the optimal zero state x_{op0} is added to the pulse pattern provided by the 3L-modulator.

Another option for the PALB is to predict the thermal behavior for the entire output period T_1 or the half of T_1 [91]. However, the number of zero state combinations for a modulation with more than 2 pulses (i.e. 4 zero state intervals) is very difficult to evaluate. As an example, for a PWM at f_{sw}=450 Hz, it is calculated as $3^{zs}=3^9=19683$ zero state combinations.

Another investigated cost function used as variables the junction temperatures at the time instant just before the next commutation, i.e. at $k+n$. However, the obtained results were not better compared to the ALB method. The reason is that the junction temperature reaches a maximum during $t_{prediction}$ and not necessarily at the end of the time interval.

5.2 Electrical and thermal modeling

Two power semiconductors are going to be evaluated in this thesis. The evaluated examples are a 3.3 kV 3L-ANPC VSC using 4.5 kV PP IGBTs and the 3.3 kV 3L-ANPC VSC using 4.5 kV IGBT modules.

5. PALB method for 3L-ANPC VSC

Table 5.1 summarizes the electrical parameters of the selected power converters.

Table 5.1 Electrical parameters of the selected power converters

Output line to line voltage	3.3 kV	3.3 kV
Semiconductor devices	CM900HB90H	T1200EB45E D1031SH45T
Nominal dc-link voltage $U_{dc,n}$	4854 V	5000 V
Commutation voltage U_{comm}	2427 V	2500 V
Maximum device voltage U_{DRM}	4500 V	4500 V
Nominal current $I_{c,n}$	900 A	1200 A

The equations of the losses are described by means of load current and voltage. In this thesis, the power losses are calculated at 125 °C. The switching loss energies of the 4.5 kV Westcode PP IGBT and the 4.5 kV Infineon PP diode are obtained experimentally. Details about their characterization can be found in [92] and [93]. The switching loss energies are calculated as:

$$\frac{E_{sw}}{\text{Ws}}\left(i_{ph}, U_{comm}\right) = \frac{U_{comm}}{U_{base}} \cdot \left(A_{sw} \cdot \left(\frac{i_{ph}}{\text{A}}\right)^2 + B_{sw} \cdot \frac{i_{ph}}{\text{A}} + C_{sw}\right), \qquad (5.6)$$

where i_{ph} is the phase current, U_{comm} is the commutation voltage and U_{base} is the voltage at which the losses were measured. The terms A_{sw}, B_{sw}, and C_{sw} denote the fitting constants obtained from the measured data at 125°C.

The conduction losses of the PP IGBT and the PP diode are calculated as:

$$P_{cond}\left(i_{ph}\right) = i_{ph} \cdot \left(V_{\Gamma 0} + r_{\Gamma 0} \cdot i_{ph}\right) \qquad (5.7)$$

where $V_{\Gamma 0}$ is the threshold voltage and $r_{\Gamma 0}$ is the slope resistance. These parameters are taken from the datasheet.

The loss approximations of the 4.5 kV Mitsubishi CM900HB90H IGBT module are taken from the datasheet [74], [94]. The switching loss energies are calculated as:

80

$$\frac{E_{sw}}{Ws}\left(i_{ph}, U_{comm}\right) = \frac{U_{comm}}{U_{base}} \cdot \left(A_{sw} \cdot \left(\frac{i_{ph}}{A}\right)^{B_{sw}}\right), \tag{5.8}$$

where A_{sw} and B_{sw} are the fitting constants.

The conduction losses for the above mentioned IGBT modules are calculated as:

$$P_{cond}\left(i_{ph}\right) = i_{ph} \cdot \left(V_{\Gamma 0} + r_{\Gamma 0} \cdot i_{ph}^{B_{cond}}\right), \tag{5.9}$$

where $V_{\Gamma 0}$ is the threshold voltage, $r_{\Gamma 0}$ is the slope resistance and B_{cond} is the fitting constant.

The switching losses are calculated as instantaneous values by averaging the sampled switching loss energy over the sample time (see Figure 5.2):

$$P_{sw} = E_{sw}\left(i_{ph}, U_{comm}\right) \cdot \frac{1}{T_{sample}}, \tag{5.10}$$

where E_{sw} is the switching loss energy, i_{ph} is the phase current, U_{comm} is the commutation voltage and T_{sample} is the sample time.

Table 5.2 Parameters and fitting constants of the selected power semiconductors

		Energy in Ws	A_{sw}	B_{sw}	C_{sw}	$V_{\Gamma 0}$ in V	$r_{\Gamma 0}$ in mΩ	B_{cond}
Westcode T1200EB45E PP IGBT and Infineon D1031SH45T PP diode								
IGBT	E_{on}	$1.616 \cdot 10^{-7}$	0.002800	0.2481	1.8	$1.6 \cdot 10^{-3}$	
	E_{off}	$-4.973 \cdot 10^{-7}$	0.004535	0.2177			
Diode	E_{rec}	$-4.716 \cdot 10^{-7}$	0.002829	0.5204	1.705	$0.928 \cdot 10^{-3}$	
Mitsubishi CM900HB90H IGBT module								
IGBT	E_{on}	0.00389	1.01645		1.2	0.01861	0.68985
	E_{off}	0.06217	0.55766				
Diode	E_{rec}	0.02859	0.48915		0.8	0.023196	0.70082

5. PALB method for 3L-ANPC VSC

The parameters and fitting constants of the Westcode T1200EB45E PP IGBT, the Infineon D1031SH45T PP diode, and the Mitsubishi CM900HB90H IGBT module are summarized in Table 5.2.

The estimated losses are fed into the semiconductor's thermal model, which is described by the Foster thermal network (see Section 4.2.2). Table 5.3 summarizes the power semiconductors and thermal parameters R_{th} and τ_{th} that can be found in datasheets [94], [95]. For the T1200EB45E Westcode PP IGBT, the thermal parameters are extracted from the transient thermal impedance shown in data sheet [94] using the method presented in [85]. A water cooling heat sink is used to dissipate the power semiconductors heat. The PP IGBT and the PP diode are installed on separate heat sinks (see thermal network from Figure 4.4 (a)). For the Mitsubishi module, the IGBT and the diode share a common heat sink, where $R_{th,ca}=13$ mK/W and $\tau_{th,ca}=2$ s (see thermal network from Figure 4.4 (b)).

Table 5.3 Thermal coefficients of the power semiconductors

	Position n	1	2	3	4	5	Heat sink
	Westcode T1200EB45E PP IGBT and Infineon D1031SH45T PP diode						
IGBT	$R_{th,n,T}$ in mK/W	3.40	2.97	1.62			6.98
	$\tau_{th,n,T}$ in s	0.55429	0.09817	0.00337			5.49
Diode	$R_{th,n,D}$ in mK/W	3.54	3.73	1.55	1.55	0.25	7.48
	$\tau_{th,n,D}$ in s	0.9	0.118	0.0282	0.00422	0.00134	5.49
	Mitsubishi CM900HB90H IGBT module						
IGBT	$R_{th,n,T}$ in mK/W	2.4	0.6	2.5	4.5		common heat sink $R_{th,ca}=13$ $\tau_{th,ca}=2$
	$\tau_{th,n,T}$ in s	1	0.3	0.1	0.03		
Diode	$R_{th,n,D}$ in mK/W	4.8	1.2	5	9		
	$\tau_{th,n,D}$ in s	1	0.3	0.1	0.03		

5.3 Comparison of the PALB and the ALB methods

The PALB and the ALB methods are compared regarding the loss and junction temperature distribution. The efficiency of both methods is evaluated at the converter's critical operating points defined in Table 3.5. The thermal behavior of the 3L-ANPC VSC is analyzed at grid frequency, at low frequency, and at zero speed.

The applied modulation strategy at grid frequency is PWM (3L-SVM and 2L-SVM) and SHE-PWM. At low output frequency and zero speed, the thermal behavior is examined only for 2L-SVM since this modulation technique delivers a better loss and junction temperature distribution. The SHE is not suitable at low output frequencies due to a large number of switching angles.

To evaluate the performance of the balancing methods, three operation modes for the 3L-ANPC VSC are considered: the NPC equivalent, applying the ALB method, and applying the PALB method. The aforementioned operation modes are compared using the same modulation pattern. The analysis of the converter's thermal behavior has shown that there is a noticeable difference between the average and the maximal junction temperature. Therefore, both thermal factors are investigated. The junction temperature ripple will be considered only at low frequency. The decrease of the maximal junction temperature achieved by the ALB and the PALB method are defined as:

$$p_{max\ ALB}(\%) = \frac{\vartheta_{j,max\ NPC} - \vartheta_{j,max\ ALB}}{\vartheta_{j,max\ NPC}} \cdot 100 \, , \qquad (5.15)$$

$$p_{max\ PALB}(\%) = \frac{\vartheta_{j,max\ NPC} - \vartheta_{j,max\ PALB}}{\vartheta_{j,max\ NPC}} \cdot 100 \, , \qquad (5.16)$$

where $\vartheta_{j,max\ NPC/ALB/PALB} = \max(\vartheta_{j,max\_T1}, \cdots, \vartheta_{jmax\_D6})_{NPC/ALB/PALB}$.

$\vartheta_{j,max\ NPC}$, $\vartheta_{j,max\ ALB}$ and $\vartheta_{j,max\ PALB}$ are the maximal junction temperatures from the NPC, the ALB, and the PALB operation mode, respectively.

5. PALB method for 3L-ANPC VSC

Similarly, the decrease of the average junction temperature applying the ALB and the PALB method are defined as:

$$P_{avg\ ALB}(\%) = \frac{\vartheta_{j,avg\ NPC} - \vartheta_{j,avg\ ALB}}{\vartheta_{j,avg\ NPC}} \cdot 100 , \tag{5.17}$$

$$P_{avg\ PALB}(\%) = \frac{\vartheta_{j,avg\ NPC} - \vartheta_{j,avg\ PALB}}{\vartheta_{j,avg\ NPC}} \cdot 100 , \tag{5.18}$$

where $\vartheta_{j,avg\ NPC/ALB/PALB} = \max(\vartheta_{j,avg\_T1}, \cdots, \vartheta_{j,avg\_D6})_{NPC/ALB/PALB} \cdot \vartheta_{j,avg\ NPC}$, $\vartheta_{j,avg\ ALB}$, and $\vartheta_{j,avg\ PALB}$ are the maximum average junction temperatures from the NPC, the ALB, and the PALB operation mode, respectively.

5.3.1 Operation at grid frequency

5.3.1.1 Evaluation of a 3.3 kV 3L-ANPC VSC using 4.5 kV press-pack IGBTs

The performances of a 3.3 kV 3L-ANPC VSC using 4.5 kV PP IGBTs applying the PALB method is evaluated at grid frequency. Table 5.4 summarizes the simulations parameters. The results are evaluated in comparison to the ALB method.

Table 5.4 Simulation parameters for a 3.3 kV 3L-ANPC VSC using 4.5 kV PP IGBTs

Operating points				Simulation parameters of a 3.3 kV 3L-NPC VSC 4.5 kV PP Westcode T1200EB45E IGBTs and 4.5 kV PP Infineon D1031SH45T diodes					
Case	pf	m_a	PWM scheme	f_{sw} in Hz	$i_{ph,rms}$ in A	f_l in Hz	U_{dc} in V	ϑ_a in °C	$\vartheta_{j,max}$ in °C
A	+1	1.15	3L-SVM						
B	−1	1.15	3L-SVM	450	1200	50	5000	50	125
C	±1	0.05	2L-SVM						

Figure 5.3 shows the loss and junction temperature distribution in **case A**. In the NPC operation mode, the outer switches T_{out} are the most stressed devices. Applying the PALB and the ALB method, respectively, the switching losses of T_{out} are equally distributed between T_{out} and T_{in}, achieving a junction temperature balance. Thus, the junction temperature of T_{out} is decreased and the junction temperature of T_{in} is increased. The maximal junction temperature is reduced from 115.9 °C to 102.4 °C by 11.6% applying the ALB method and to 101.9 °C by 12.1% applying the PALB method (see Table 5.5). Also, both balancing methods distribute the switching and conduction losses of D_{NPC} between D_{in} and D_{NPC}.

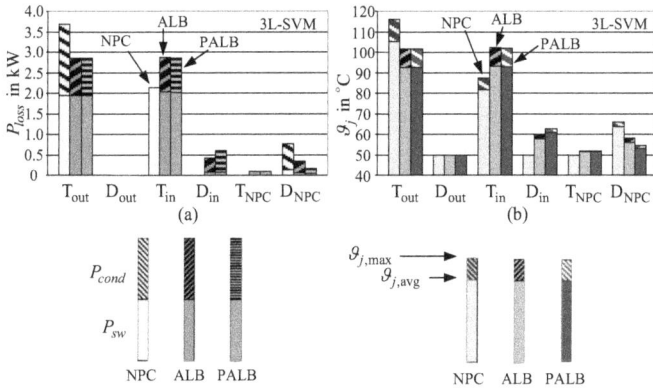

Figure 5.3 (a) Loss and (b) junction temperature distribution in a 3.3 kV 3L-ANPC VSC using 4.5 kV PP IGBTs, U_{dc}=5000 V, $i_{ph,rms}$=1200 A, f_{sw}=450 Hz, f_1=50 Hz, $\vartheta_a = 50$ °C, pf=1, m_a=1.15, 3L-SVM

Figure 5.4 shows the critical junction temperatures of T_{in}, the phase current, and the zero states selection applying the balancing methods in case A. Since the inner switches T_2 and T_3 have a symmetrical thermal behavior for positive and negative phase currents, only the critical junction temperature of T_2 is illustrated. Figure 5.4 (a) shows the junction temperature waveforms and the average junction temperature of the inner switch T_2 applying the balancing methods. Both balancing methods reduce the junction temperature ripple from 17 to 14 K.

5. PALB method for 3L-ANPC VSC

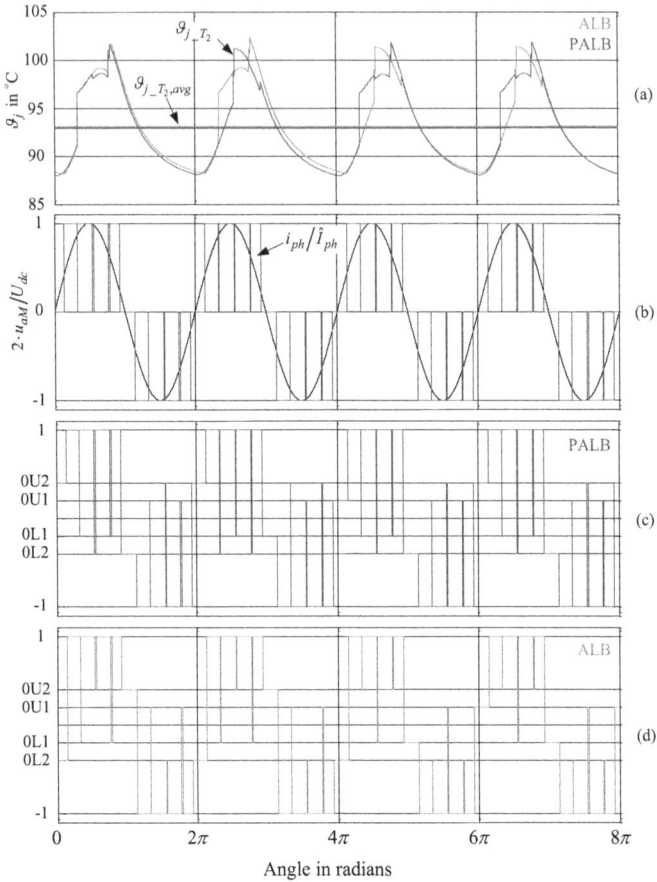

Figure 5.4 Junction temperature waveforms in a 3.3 kV 3L-ANPC VSC using 4.5 kV PP IGBTs, (a) Junction temperature of T₂; (b) Normalized phase-midpoint output voltage and normalized phase current; (c) Zero states selection applying the PALB method; (d) Zero states selection applying the ALB method; U_{dc}=5000 V, $i_{ph,rms}$=1200 A, f_{sw}=450 Hz, f_1=50 Hz, ϑ_a = 50 °C, pf=1, m_a=1.15, 3L-SVM

Figure 5.4 (b) shows the normalized phase-midpoint output voltage and normalized phase current. Figures 5.4 (c) and Figures 5.4 (d) show the selection of the zeros states applying the ALB and the PALB

86

method, respectively. The pulse pattern of the PALB method repeats every 8 output periods, whereas the pulse pattern of the ALB method repeats every 6 output periods. Since the objective of Figure 5.4 is to emphasize the different selection of the zero states applying the ALB and the PALB methods, only 4 output periods are shown. The principle of operation of the two balancing methods is evaluated regarding the selection of zero states and their equivalent types of commutations.

Figure 5.5 illustrates the comparison of the commutation types used by the two balancing methods in case A. The utilization percentage of the Type 3 commutation is 50% for both balancing methods. The difference is that the PALB method uses 50% Type 2 commutations, whereas the ALB method uses 50% Type 1 commutation.

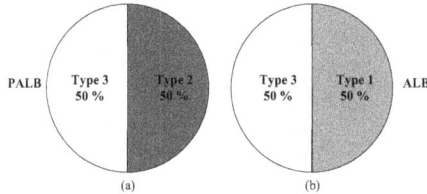

(a)   (b)

Figure 5.5 Commutations types in a 3.3 kV 3L-ANPC VSC using 4.5 kV PP IGBTs applying (a) the PALB method, (b) the ALB method; U_{dc}=5000 V, $i_{ph,rms}$=1200 A, f_{sw}=450 Hz, f_1=50 Hz, $\vartheta_a = 50$ °C, pf=1, m_a=1.15, 3L-SVM

The PALB method selects the Type 2 commutation to decrease the junction temperature of T_{in} and the Type 3 commutations to decrease the junction temperature of T_{out}. The combination of Type 2 and Type 3 commutations accomplish a symmetrical distribution of the switching losses between T_{out} and T_{in}. The PALB chooses the Type 2 commutation and transfers conduction losses during the zero state from T_{in} to T_{NPC}. However, these conduction losses are small compared with the total losses of T_{in}. Consequently, the maximal junction temperature is reduced only by 0.5% applying the PALB method compared to the ALB method. On the other hand, the selection of the Type 2 commutation increases the imbalance between D_{NPC} and D_{in}. The ALB method uses a combination of Type 1 and Type 3 commutations to balance the junction temperature between T_{in} and T_{out} and between D_{in} and D_{NPC}. Since the main goal of

87

5. PALB method for 3L-ANPC VSC

the ALB method is thermal balancing, the Type 2 commutation is not required.

Figure 5.6 shows the loss and junction temperature distribution in **case B**. The most stressed devices are the outer diodes D_{out}, in all three operation modes, NPC, PALB, and ALB. Applying the balancing methods, the switching losses of D_{out} are distributed between D_{out} and D_{in}. Thus, a thermal balance between D_{out} and D_{in} is achieved. Both balancing methods reduce the maximal junction temperature of D_{out} from 91.2 °C to 85.9 °C by 5.6%. Thus, no further improvement is made by the PALB method.

Both balancing methods distribute the switching losses of T_{in} between T_{in} and T_{NPC}. The ALB method accomplishes a better thermal balance between T_{in} and T_{NPC}.

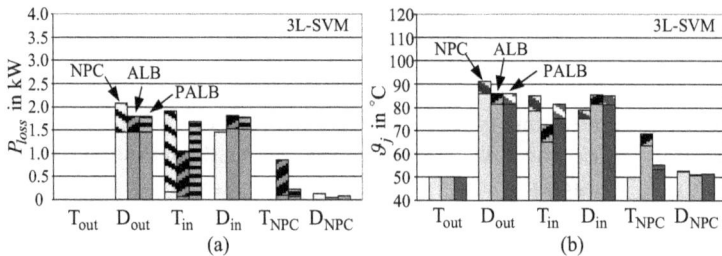

Figure 5.6 (a) Loss and (b) junction temperature distribution in a 3.3 kV 3L-ANPC VSC using 4.5 kV PP IGBTs U_{dc}=5000 V, $i_{ph,rms}$=1200 A, f_{sw}=450 Hz, f_1=50 Hz, ϑ_a = 50 °C, pf=-1, m_a=1.15, 3L-SVM

Figure 5.7 shows in comparison the critical junction temperatures of D_{out}, the phase current, and the zero states selection in case B. Since the outer diodes D_1 and D_4 have a symmetrical thermal behavior for positive and negative phase currents, only the critical junction temperature of D_1 is illustrated. Figure 5.7 (a) shows the junction temperature waveform and the average junction temperature of the outer diode D_1 applying the balancing methods. The junction temperature ripple is less than 10 K und thus, uncritical [9].

88

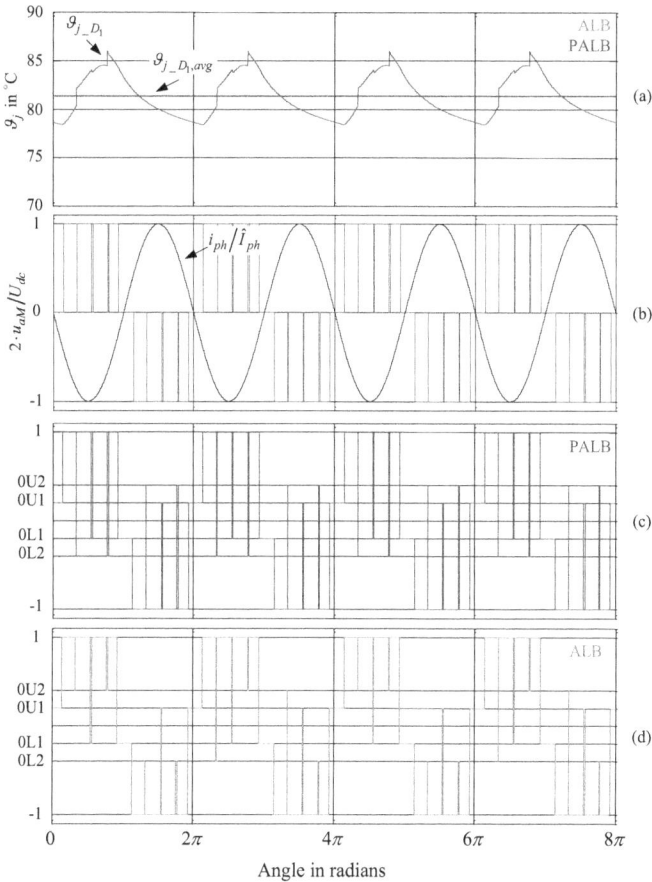

Figure 5.7 Junction temperature waveforms in a 3.3 kV 3L-ANPC VSC using 4.5 kV PP IGBTs, (a) Junction temperature of D_1; (b) Normalized phase midpoint output voltage and normalized phase current; (c) Zero states selection applying the PALB method; (d) Zero states selection applying the ALB method; U_{dc}=5000 V, $i_{ph,rms}$=1200 A, f_{sw}=450 Hz, f_1=50 Hz, ϑ_a = 50 °C, pf=-1, m_a=1.15, 3L-SVM

Figure 5.7 (b) shows the normalized phase-midpoint output voltage and normalized phase current. Also in case B, the ALB and the PALB methods present a different selection of the zero states (see Figure

89

5. PALB method for 3L-ANPC VSC

5.7 (c) and Figure 5.7 (d)). The pulse pattern of the ALB method repeats every 2 output periods, whereas the pulse pattern of the PALB method remains the same.

Figure 5.8 illustrates the comparison of the commutation types used by the two balancing methods in case B. The utilization percentage of the Type 2 commutation is 50% applying the PALB method and 12% applying the ALB method. Both balancing methods present 50% utilization percentage of the Type 3 commutation. Thus, using a combination of Type 2 and Type 3 commutations, both methods achieve a thermal balance between D_{out} and D_{in} and a decrease of the maximal junction temperature of D_{out}.

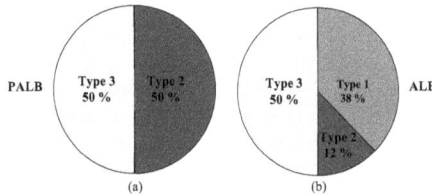

(a)   (b)

Figure 5.8 Commutations types in a 3.3 kV 3L-ANPC VSC using 4.5 kV PP IGBTs applying; (a) the PALB; and (b) the ALB method; U_{dc}=5000 V, $i_{ph,rms}$=1200 A, f_{sw}=450 Hz, f_1=50 Hz, ϑ_a = 50 °C, pf=-1, m_a=1.15, 3L-SVM

The difference between the two balancing methods is that the ALB method presents 38% utilization of the Type 1 commutation. Thus, using a combination of Type 1 and Type 2 commutations, a better thermal balance between T_{NPC} and T_{in} is achieved.

Figure 5.9 shows the loss and junction temperature distribution in **case C**. In the NPC operation mode, the inner switches T_{in} are the most stressed devices. Applying the PALB and the ALB method, the switching and conduction losses of T_{in} are distributed between T_{in}, T_{out}, and T_{NPC}. Thus, an improved thermal balance is achieved: the junction temperature of T_{in} is decreased while the junction temperature T_{NPC} is increased.

Applying the balancing methods, the outer switches T_{out} present the highest junction temperature. The ALB method reduces the maximal

90

junction temperature from 97.5 °C to 81.9 °C by 15.9 %. Applying the PALB method, the maximal junction temperature is reduced from 97.5 °C to 78.7 °C by 19.2%. Thus, the PALB method achieves an improvement of 3.3% compared to the ALB method (see Table 5.5).

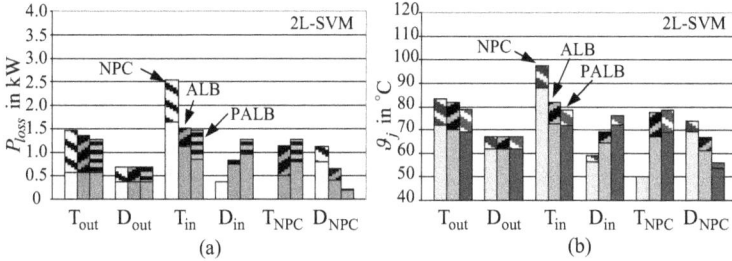

Figure 5.9 (a) Loss and (b) junction temperature distribution in a 3.3 kV 3L-ANPC VSC using 4.5 kV PP IGBTs U_{dc}=5000 V, $i_{ph,rms}$=1200 A, f_{sw}=450 Hz, f_1=50 Hz, ϑ_a = 50 °C, pf=±1, m_a=1.15, 3L-SVM

Figure 5.10 shows the critical junction temperatures of T_{out}, the phase current, and the zero states selection applying the balancing methods in case C. Since the outer switches T_1 and T_4 have a symmetrical thermal behavior for positive and negative phase currents, only the critical junction temperature of T_1 is illustrated. Figure 5.10 (a) shows the junction temperature waveform and the average junction temperature of the outer diode T_1 applying the balancing methods. Although the maximal junction temperature was reduced applying the balancing methods, the junction temperature ripple remains approximately 15 K. Figure 5.10 (b) shows the normalized phase-midpoint output voltage and normalized phase current. For the 2L-SVM, the reference waveforms are shifted into the upper or lower carrier band by a common-mode dc-offset that changes its polarity with the half of the fundamental frequency. Thus, there are 4 sectors identified for this operation point: Sector 1: $u_{aM}>0$ & $i_{ph}>0$; Sector 2: $u_{aM}>0$ & $i_{ph}<0$; Sector 3: $u_{aM}<0$ & $i_{ph}>0$; Sector 4: $u_{aM}<0$ & $i_{ph}<0$.

91

5. PALB method for 3L-ANPC VSC

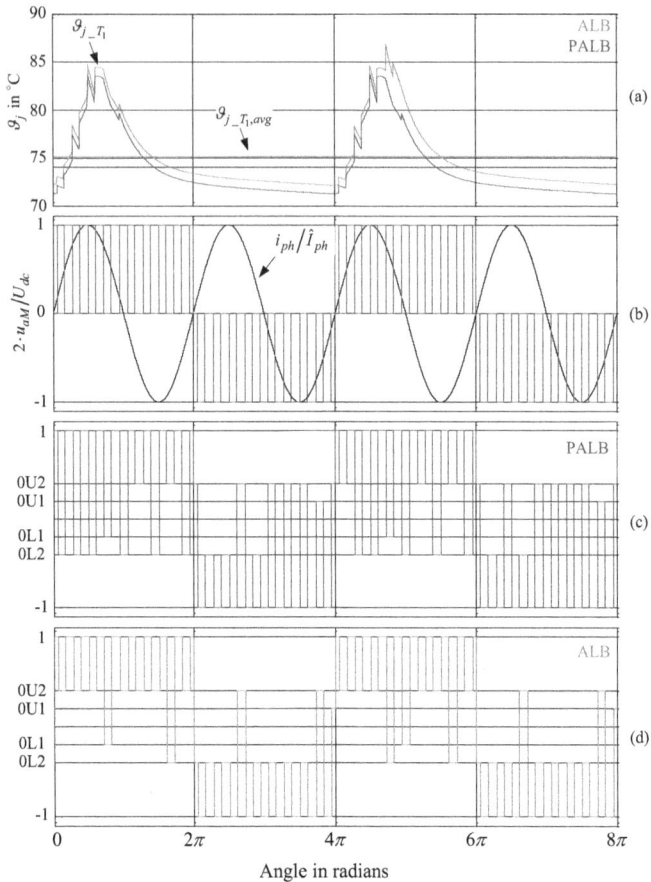

Figure 5.10 Junction temperature waveforms in a 3.3 kV 3L-ANPC VSC using 4.5 kV PP IGBTs, (a) Junction temperature of T_1; (b) Normalized phase midpoint output voltage and normalized phase current; (c) Zero states selection applying the PALB method; (d) Zero states selection applying the ALB method; U_{dc}=5000 V, $i_{ph,rms}$=1200 A, f_{sw}=450 Hz, f_1=50 Hz, ϑ_a = 50 °C, pf=1, m_a=0.05, 2L-SVM

The commutation types used are symmetrical during Sector 1 and 4 (analog to inverter operation for one phase leg) and Sector 2 and 3 (analog to rectifier operation for one phase leg). Thus, the commutation

92

types used by the balancing methods are analyzed during Sector 1 and 4 and during Sector 2 and 3. Also, in case C, the ALB and the PALB methods present a different selection of the zero states (see Figure 5.10 (c) and Figure 5.10 (d)).

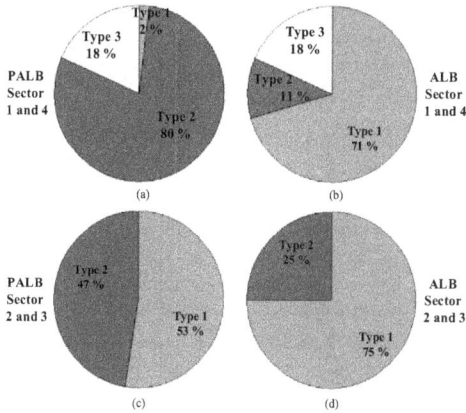

Figure 5.11 Commutations types in a 3.3 kV 3L-ANPC VSC using 4.5 kV PP IGBTs applying (a) PALB method in Sector 1 and 4; (b) ALB method in Sector 1 and 4; (c) PALB method in Sector 2 and 3; (d) ALB method in Sector 2 and 3; U_{dc}=5000 V, $i_{ph,rms}$=1200 A, f_{sw}=450 Hz, f_1=50 Hz, ϑ_a = 50 °C, pf=1, m_a=0.05, 2L-SVM

Figure 5.11 shows the commutation types used by the balancing methods in case C. During Sector 1/4, the PALB method uses a combination of Type 2 and Type 3 commutations and only 2% of Type 1 commutation. The PALB method presents a utilization of 80% of Type 2 commutation to reduce the critical junction temperature of T_{in}. As a result, the maximal junction temperature is decreased by 3.3% compared to the ALB method. On the other hand, the ALB method uses a combination of Type 1 and Type 3 commutations to balance the junction temperature between T_{in} and T_{out} and between D_{in} and D_{NPC}. Only a low percentage of the Type 2 commutation, i.e. 11%, is used because this commutation increases the imbalance of the junction temperatures of D_{NPC} and D_{in}. During Sector 2/3, both balancing methods use a combination of Type 1 and Type 2 commutations to balance the junction temperature between

93

the T_{in} and T_{NPC}. The difference is that the PALB method selects more often the Type 2 commutation to reduce the maximal junction temperature of T_{NPC}.

Table 5.5 summarizes the decrease of the maximal and average junction temperature applying the balancing methods. The PALB method presents a slightly better performance compared to the ALB method.

Table 5.5 Junction temperature decrease applying the PALB and the ALB methods in a 3.3 kV 3L-ANPC VSC using 4.5 kV PP IGBTs

Decrease of the maximal junction temperature ϑ_{max}			
Case	$p_{max\ ALB}$ in %	$p_{max\ PALB}$ in %	$p_{max\ PALB}$- $p_{max\ ALB}$ in %
A	11.6	12.1	0.5
B	5.6	5.6	0
C	15.9	19.2	3.3
Decrease of the average junction temperature ϑ_{avg}			
	$P_{avg\ ALB}$ in %	$P_{avg\ PALB}$ in %	$P_{avg\ PALB}$- $p_{avg\ ALB}$ in %
A	11.4	11.7	0.3
B	5.5	5.5	0
C	17.3	17.8	0.5

In the following paragraph, the performance of the balancing methods is investigated for the 3L-ANPC VSC applying selective harmonic elimination using 5 angles (SHE-5α). Table 5.6 summarizes the simulations parameters.

The NPC mode is characterized by the maximal junction temperature of T_{out} in **case A** (see Figure 5.12 (a) and Figure 5.13 (a)). The ALB method uses a combination of 38% of Type 1 commutations, 14% of Type 2 commutations, and 48% of Type 3 commutations, achieving thermal balance between T_{in} and T_{out} and between D_{NPC} and D_{in}. Thus, the maximal junction temperature is reduced from 111.5 °C to 100.5 °C by 9.9% (see Table 5.7). The PALB method uses a combination of 44% of Type 2 commutations and 56% of Type 3 commutations, achieving a thermal balance between T_{in} and T_{out}. Thus, the maximal junction temperature is reduced from 111.5 °C to 100.2 °C by 10.1% (see

94

Table 5.7). The PALB has an improvement of 0.2%. On the other hand, the higher utilization of the Type 2 commutation causes an imbalance between D_{NPC} and D_{in}.

Table 5.6 Simulation parameters for a 3.3 kV 3L-ANPC VSC using 4.5 kV PP IGBTs

Operating points				Simulation parameters of a 3.3 kV 3L-NPC VSC 4.5 kV PP Westcode T1200EB45E IGBTs and 4.5 kV PP Infineon D1031SH45T diodes					
Case	pf	m_a	PWM scheme	f_{sw} in Hz	$i_{ph,rms}$ in A	f_1 in Hz	U_{dc} in V	ϑ_a in °C	$\vartheta_{j,max}$ in °C
A	+1	1.15	SHE-5α						
B	−1	1.15	SHE-5α		1200				
C	+1	0.05	SHE-5α	500		50	5000	50	125
D	−1	0.05	SHE-5α		1060				
E	±1	0.05	2L-SVM		1200				

The outer diodes D_{out} represent the critical devices regarding the maximal junction temperature in **case B** (see Figure 5.12 (b) and Figure 5.13 (b)). The ALB method uses a combination of 52% of Type 1 commutations and 48% of Type 3 commutations, achieving thermal balance between D_{out} and D_{in} and between T_{in} and T_{NPC}. Thus, the maximal junction temperature is reduced from 91.2 °C to 86.2 °C by 5.5%. The PALB method uses a combination of 56% of Type 2 commutations and 44% of Type 3 commutations, achieving thermal balance between D_{out} and D_{in}. Thus, the PALB method decreases the maximal junction temperature from 91.2 °C to 85.8 °C by 5.9 % (see Table 5.7).

Case C (see Figure 5.12 (c) and Figure 5.13 (c)) is characterized by the highest junction temperature of T_{out}. Both balancing methods use a combination of all three commutation types. The ALB method uses 44% of Type 1, 16% of Type 2 and 40% of Type 3 commutations. The ALB method achieves junction temperature balance between T_{out} and T_{in} and between D_{in} and D_{NPC}. Thus, the ALB method reduces the maximal junction temperature of T_{out} from 102.8 °C to 87.8 °C by 14.6%. The

5. PALB method for 3L-ANPC VSC

PALB method uses 21% of Type 1, 39% of Type 2 and 40% of Type 3 commutations. The PALB method reduces the maximal junction temperature from 102.8 °C to 85.2 °C by 17.2% (see Table 5.7). Thus, an improvement of 2.6% is achieved applying the PALB method compared to the ALB method.

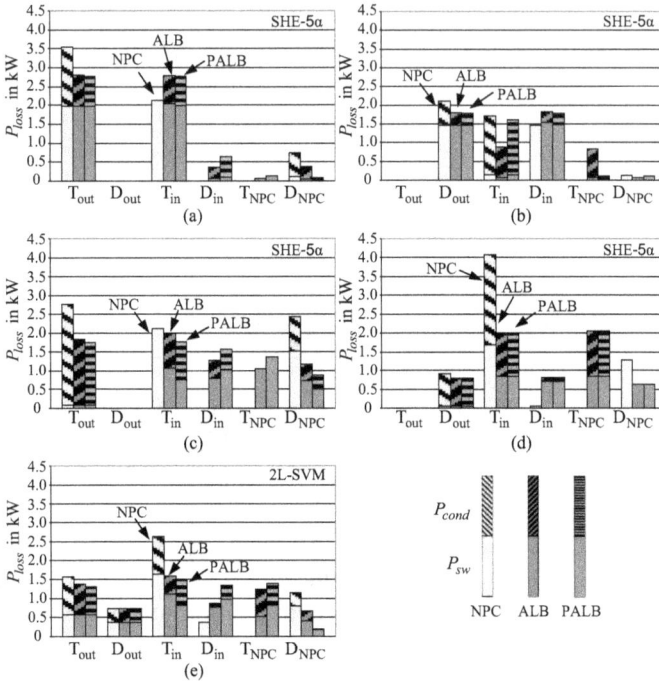

Figure 5.12 Loss distribution in a 3.3 kV 3L-ANPC VSC using 4.5 kV PP IGBTs; (a) pf=1, m_a=1.15; (b) pf=-1, m_a=1.15; (c) pf=1, m_a=0.05; (d) pf=-1, m_a=0.05, $i_{ph,rms}$=1060 A; (e) pf=±1, m_a=0.05, 2L-SVM; U_{dc}=5000 V, $i_{ph,rms}$=1200 A, f_{sw}=500 Hz, f_1=50 Hz, $\vartheta_a = 50°C$, SHE-5α

In **case D** (see Figure 5.12 (d) and Figure 5.13 (d)), because the maximal allowable junction temperature is 125 °C, the phase current has to be set to 1060 A (see Table 5.6). The inner switches T_{in} present the maximal junction temperature in the NPC operation mode. Applying the balancing methods, the maximal junction temperature is reached in T_{NPC}.

96

Both balancing methods use a combination of all three types of commutation with similar percentage. The ALB method uses 40% of Type 1, 44% of Type 2 and 16% of Type 3 commutations, whereas the PALB method uses 40% of Type 1, 48% of Type 2 and 12% of Type 3 commutations. Both balancing methods achieve thermal equilibrium between D_{out} and D_{in} and between T_{in} and T_{NPC}. The maximal junction temperature is reduced from 124.5 °C to 88.9 °C by 28.5% (see Table 5.7).

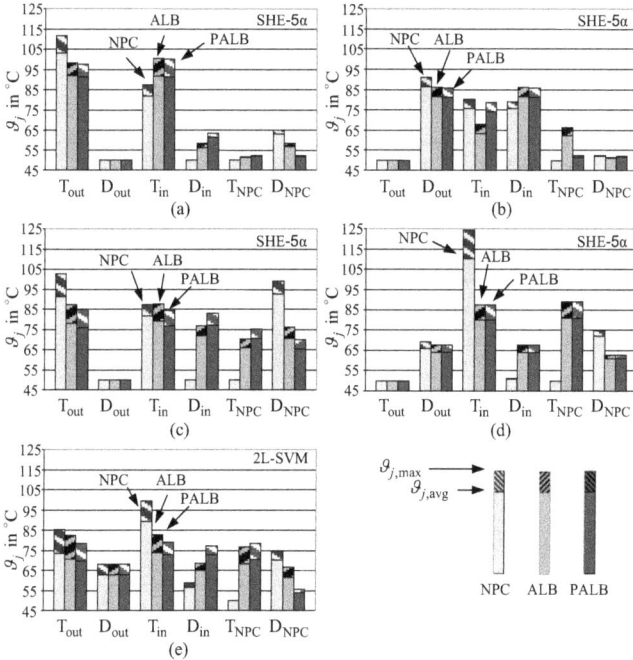

Figure 5.13 Junction temperature distribution in a 3.3 kV 3L-ANPC VSC using 4.5 kV PP IGBTs, (a) pf=1, m_a=1.15; (b) pf=-1, m_a=1.15; (c) pf=1, m_a=0.05; (d) pf=-1, m_a=0.05, $i_{ph,rms}$=1060 A; (e) pf=±1, m_a=0.05, 2L-SVM; U_{dc}=5000 V, $i_{ph,rms}$=1200 A, f_{sw}=500 Hz, f_1=50 Hz, ϑ_a = 50°C, SHE-5α

The inner switches T_{in} represent the critical devices regarding the maximal junction temperature in **case E** (see Figure 5.12 (e) and Figure 5.13 (e)). Both balancing methods use all three types of commutations.

97

5. PALB method for 3L-ANPC VSC

The ALB method reduces the maximal junction temperature of T_{in} from 97.5 °C to 81.9 °C by 16.6%, whereas the PALB method reduces the maximal junction temperature 97.5 °C to 78.7 °C by 20.2% (see Table 5.7). The PALB method features an improvement of 3.6% compared to the ALB method.

Table 5.7 summarizes the decrease of the maximal and average junction temperature applying the ALB and the PALB method, respectively.

Table 5.7 Junction temperature decrease applying the PALB and the ALB methods in a 3.3 kV 3L-ANPC VSC using 4.5 kV PP IGBTs

	Decrease of the maximal junction temperature ϑ_{max}		
Case	$p_{max\ ALB}$ in %	$p_{max\ PALB}$ in %	$p_{max\ PALB}$- $p_{max\ ALB}$ in %
A	9.9	10.1	0.2
B	5.5	5.9	0.4
C	14.6	17.2	2.6
D	28.5	28.5	0
E	16.6	20.2	3.6
	Decrease of the average junction temperature ϑ_{avg}		
	$P_{avg\ ALB}$ in %	$P_{avg\ PALB}$ in %	$P_{avg\ PALB}$- $p_{avg\ ALB}$ in %
A	10.7	11.3	0.6
B	5.6	6.1	0.5
C	14.6	16.5	1.9
D	26.7	26.7	0
E	17.4	18.4	1.0

In **Appendix A**, the performance of the balancing methods is further evaluated at f_{sw}=300 Hz (SHE-3α), f_{sw}=750 Hz (SVM), and f_{sw}=1050 Hz (SVM). The PALB method presents an improved performance compared to the ALB method regarding the decrease of the maximal junction temperature by up to 1.9%.

5.3.1.2 Evaluation of a 3.3 kV 3L-ANPC VSC using 4.5 kV IGBT modules

The performance of the balancing methods is evaluated for a 3.3 kV 3L-ANPC VSC using IGBT modules. It is considered that the converter operates at grid frequency (50 Hz). The thermal analysis of the 3L-ANPC VSC using IGBT modules empathize the influence of the thermal coupling between the power switch and the diode in one module. Due to thermal reliability of the IGBT module, the maximum rise of the junction temperature should not exceed 30 K [96]. Thus, for a cooling water temperature of $\vartheta_a = 37$ °C, the maximum average junction temperature is limited to 67 °C. Table 5.8 summarizes the simulations parameters.

Table 5.8 Simulation parameters for a 3.3 kV 3L-ANPC VSC using 4.5 kV IGBT modules

Operating points				Simulation parameters of a 3.3 kV 3L-ANPC VSC 4.5 kV Mitsubishi CM900HB90H IGBT module					
Case	pf	m_a	PWM scheme	f_{sw} in Hz	$i_{ph,rms}$ in A	f_1 in Hz	U_{dc} in V	ϑ_a in °C	$\vartheta_{j,max}$ in °C
A	+1	1.15	3L-SVM		460				
B	−1	1.15	3L-SVM	450	460	50	4854	37	67
C	±1	0.05	2L-SVM		625				

Case A (see Figure 5.14) is characterized by the maximal junction temperature of T_{out} in the NPC mode. The PALB and the ALB methods accomplish a thermal balance between T_{out} and T_{in}. Both balancing methods reduce the maximal junction temperature from 68.8 °C to 58.9 °C by 13% (see Table 5.9).

5. PALB method for 3L-ANPC VSC

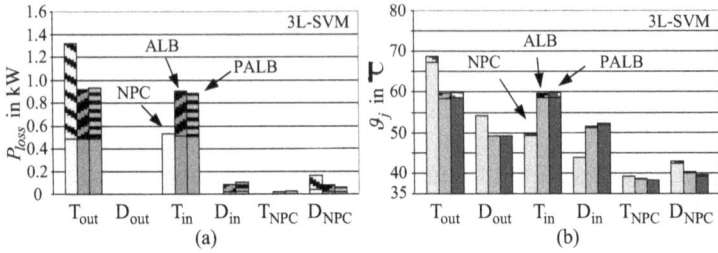

Figure 5.14 (a) Loss and (b) junction temperature distribution in a 3.3 kV 3L-ANPC VSC using 4.5 kV IGBT modules; U_{dc}=4854 V, $i_{ph,rms}$=460 A, f_{sw}=450 Hz, f_1=50 Hz, $\vartheta_a = 37°C$, pf=1, m_a=1.15, 3L-SVM

Figure 5.15 shows the comparison of the commutation types used by the two balancing methods in case A. The PALB method uses all three commutation types, whereas the ALB method uses only Type 1 and Type 3 commutations. The PALB method reduces the critical junction temperature of T_{in} by selecting the Type 2 commutation instead of the Type 3 commutation. Although the Type 2 commutation releases the inner IGBTs, it stresses the inner diodes with conduction and switching losses. Since the inner diodes are thermally coupled with the inner IGBTs, the accumulated losses in the diodes will increase the junction temperature of the parallel connected IGBTs. Thus, no improvements are achieved applying the PALB method compared to the ALB method.

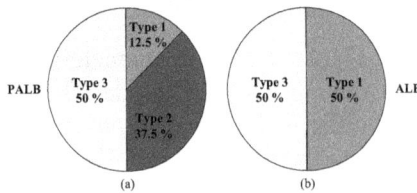

Figure 5.15 Commutations types in a 3.3 kV 3L-ANPC VSC using 4.5 kV IGBT modules applying (a) the PALB, (b) the ALB method; U_{dc}=4854 V, $i_{ph,rms}$=460 A, f_{sw}=450 Hz, f_1=50 Hz, $\vartheta_a = 37°C$, pf=1, m_a=1.15, 3L-SVM

Case B (see Figure 5.16) is characterized by the maximal junction temperature of D_{in} in all operation modes. Compared to the PALB method, the ALB method achieves a better thermal balance between T_{in}

100

and T_{NPC}. The maximal junction temperature is reduced from 65.8 °C to 57.9 °C by 10.3% applying the ALB method and from 65.8 °C to 65.1 °C by 1.1% applying the PALB method. Thus, the ALB method decreases the maximal junction temperature by 9.2% compared to the PALB method (see Table 5.9).

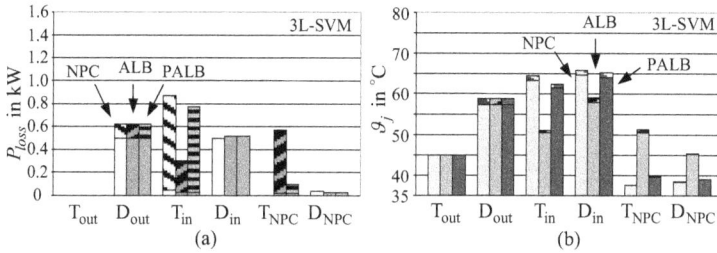

Figure 5.16 (a) Loss and (b) junction temperature distribution in a 3.3 kV 3L-ANPC VSC using 4.5 kV IGBT modules U_{dc}=4854 V, $i_{ph,rms}$=460 A, f_{sw}=450 Hz, f_1=50 Hz, $\vartheta_a = 37°C$, pf=-1, m_a=1.15, 3L-SVM

Figure 5.17 illustrates the comparison of the commutation types used by the two balancing methods in case B. The PALB method uses only Type 2 commutations. During the negative half wave of the phase current, the PALB method uses the Type 2 commutation to reduce the junction temperature of the inner diode D_2 by releasing the semiconductor from the upcoming switching losses. However, the inner switch T_3 experiences conduction and switching losses during the Type 2 commutation. Since T_3 is thermally coupled with D_3, the accumulated losses in the power switch will increase the junction temperature of D_3, which will have considerable conduction losses for the positive phase currents. The selection procedure is similar for the positive half wave of the phase current. The PALB method considers the power losses only during the prediction time up to the next commutation to a zero state and not for the entire output period. Thus, the PALB method does not take into account this effect of power loss accumulation.

5. PALB method for 3L-ANPC VSC

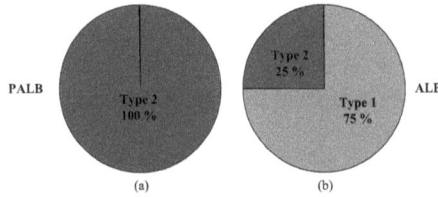

(a)   (b)

Figure 5.17 Commutations types in a 3.3 kV 3L-ANPC VSC using 4.5 kV IGBT modules applying (a) the PALB, (b) the ALB method; U_{dc}=4854 V, $i_{ph,rms}$=460 A, f_{sw}=450 Hz, f_1=50 Hz, $\vartheta_a = 37°C$, pf=-1, m_a=1.15, 3L-SVM

Applying the ALB method, the utilization percentage of the Type 1 commutation is 75%. Thus, T_{in} is stressed with a smaller amount of conduction and switching losses compared to the PALB method. Since T_{in} are thermally coupled with D_{in}, the junction temperature is lower. The use of the Type 2 commutation in a larger percentage has no advantage to reduce the highest junction temperature considering the entire fundamental period of the output frequency. Thus, the ALB method has an improvement of 9.2% compared the PALB method (see Table 5.9).

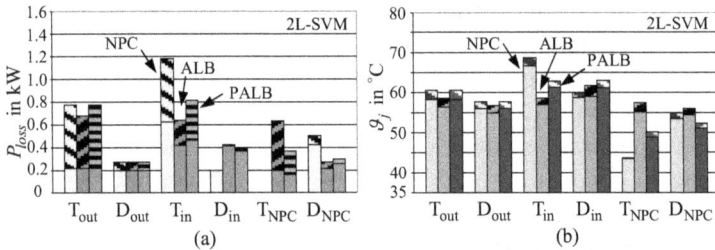

Figure 5.18 (a) Loss and (b) junction temperature distribution in a 3.3 kV 3L-ANPC VSC using 4.5 kV IGBT modules; U_{dc}=4854 V, $i_{ph,rms}$=460 A, f_{sw}=450 Hz, f_1=50 Hz, $\vartheta_a = 37°C$, pf=1, m_a=0.05, 2L-SVM

Case C (see Figure 5.18) is characterized by the maximal junction temperature of T_{in}. The balancing methods achieve thermal balancing between T_{in} and T_{out} and between D_{NPC} and D_{in}. The maximal junction temperature is reduced from 68.6 °C to 61.8 °C by 9.9% applying the ALB method and from 68.6 °C to 63.2 °C by 7.9% applying the PALB method. Thus, the ALB method has an improvement of 2% compared the PALB method (see Table 5.9).

Figure 5.19 shows the commutation types used by the balancing methods in case C. In **Sector 1 and 4**, the ALB method uses a combination of Type 1 and Type 3 commutations to achieve the thermal balance between T_{in} and T_{out} and between D_{in} and D_{NPC}. Instead of the Type 3 commutation, the PALB method chooses the Type 2 commutation in order to release the highest junction temperature of T_{in}. However, during the Type 2 commutation, D_{in} is stressed with conduction and switching losses. Since D_{in} is thermally coupled with T_{in}, the accumulated losses in the diode will increase the junction temperature of T_{in}. Summarizing, selecting the Type 2 commutation in a larger percentage presents no advantage by decreasing junction temperatures of T_{in} and D_{in}. In **Sector 2 and 3**, both balancing methods use a combination of Type 1 and Type 2 commutations to achieve thermal equilibrium between T_{in} and T_{NPC} and between D_{in} and D_{out}. However, the PALB method presents a larger percentage of Type 2 commutation with the aim to reduce the junction temperature of D_{in}. However, during the Type 2 commutation, T_{in} is stressed with conduction and switching losses. Since T_{in} is thermally coupled with D_{in}, the accumulated losses in the IGBT will increase the junction temperature of D_{in}. As a consequence, T_{in} and D_{in} present a lower junction temperature applying the ALB method compared to PALB method.

Table 5.9 summarizes the decrease of the maximal and average junction temperature applying the ALB and the PALB method, respectively.

Table 5.9 Junction temperature decrease applying the PALB and the ALB methods in a 3.3 kV 3L-ANPC VSC using 4.5 kV IGBT modules

Decrease of the maximal junction temperature ϑ_{max}			
Case	$p_{max\ ALB}$ in %	$p_{max\ PALB}$ in %	$p_{max\ PALB}$- $p_{max\ ALB}$ in %
A	13.0	12.8	-0.2
B	10.3	1.1	-9.2
C	9.9	7.9	-2

5. PALB method for 3L-ANPC VSC

	Decrease of the average junction temperature ϑ_{avg}		
	$p_{avg\,ALB}$ in %	$p_{avg\,PALB}$ in %	$p_{avg\,PALB^-}\,p_{avg\,ALB}$ in %
A	12.9	12.7	-0.2
B	10.5	1.1	-9.4
C	11.5	8.0	-3.5

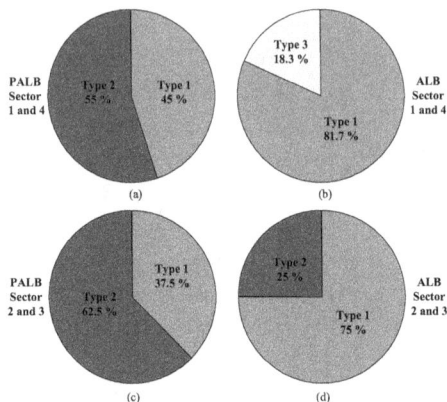

Figure 5.19 Commutations types in a 3.3 kV 3L-ANPC VSC using 4.5 kV IGBT modules applying (a) PALB method in Sector 1 and 4; (b) ALB method in Sector 1 and 4; (c) PALB in method Sector 2 and 3; (d) ALB method in Sector 2 and 3; U_{dc}=4854 V, $i_{ph,rms}$=460 A, f_{sw}=450 Hz, f_1=50 Hz, ϑ_a = 37°C, pf=1, m_a=0.05, 2L-SVM

In **Appendix B**, the performance of the balancing methods is further evaluated at f_{sw}=500 Hz (SHE-5α), f_{sw}=750 Hz (SVM), and f_{sw}=1050 Hz (SVM). The ALB method presents an improved performance compared to the PALB method regarding the decrease of the maximal junction temperature by up to 6.2%.

5.3.2 Operation at low output frequency

At low fundamental output frequencies, the junction temperature ripple increases, what negatively influences the lifetime expectancy of the

104

power device (see Section 4.2.3). Thus, to compare the performance of both balancing methods, not only the decreases of the maximal and average junction temperatures have to be considered, but also the decrease of the junction temperature ripple has to be evaluated. The decrease of the junction temperature ripple applying the ALB and the PALB method is defined as:

$$P_{\Delta \vartheta \, \text{ALB}}(\%) = \frac{\Delta \vartheta_{j \, \text{NPC}} - \Delta \vartheta_{j \, \text{ALB}}}{\Delta \vartheta_{j \, \text{NPC}}} \cdot 100 \, ,$$

(5.19)

$$P_{\Delta \vartheta \, \text{PALB}}(\%) = \frac{\Delta \vartheta_{j \, \text{NPC}} - \Delta \vartheta_{j \, \text{PALB}}}{\Delta \vartheta_{j \, \text{NPC}}} \cdot 100 \, ,$$

(5.20)

where $\Delta \vartheta_{j \, \text{NPC/ALB/PALB}} = \max(\Delta \vartheta_{j\_T1}, \dots, \Delta \vartheta_{j\_D6})_{\text{NPC/ALB/PALB}}$.

$\Delta \vartheta_{j \, \text{NPC}}$, $\Delta \vartheta_{j \, \text{ALB}}$, and $\Delta \vartheta_{j \, \text{PALB}}$ are the maximal junction temperature ripples from the NPC, the ALB, and the PALB operation mode, respectively.

5.3.2.1 Evaluation of a 3.3 kV 3L-ANPC VSC using 4.5 kV press-pack IGBTs

To determine the performance of the PALB method compared to the ALB method, the thermal behavior of the converter is evaluated at 2 Hz. Table 5.10 summarizes the simulations parameters.

Table 5.10 Simulation parameters for a 3.3 kV 3L-ANPC VSC using 4.5 kV PP IGBTs

Operating points				Simulation parameters of a 3.3 kV 3L-NPC VSC 4.5 kV PP Westcode T1200EB45E IGBTs and 4.5 kV PP Infineon D1031SH45T diodes				
pf	m_a	PWM scheme	f_{sw} in Hz	$i_{ph,rms}$ in A	f_1 in Hz	U_{dc} in V	ϑ_a in °C	$\vartheta_{j,max}$ in °C
±1	0.05	2L-SVM	450	1200	2	5000	50	125

5. PALB method for 3L-ANPC VSC

Figure 5.20 shows the loss and the junction temperature distribution of a 3.3 kV 3L-ANPC VSC at low output frequency. The inner switches are the most stressed devices in all operation modes. Applying the balancing methods, thermal equilibrium is achieved between T_{in}, T_{out}, and T_{NPC}. The maximal junction temperature is reduced from 111.9 °C to 94.2 °C by 15.8% applying the ALB method and from 111.9 °C to 90.1 °C by 19.5% applying the PALB method. Thus, the PALB method has an improvement of 3.7% compared to the ALB method (see Table 5.11).

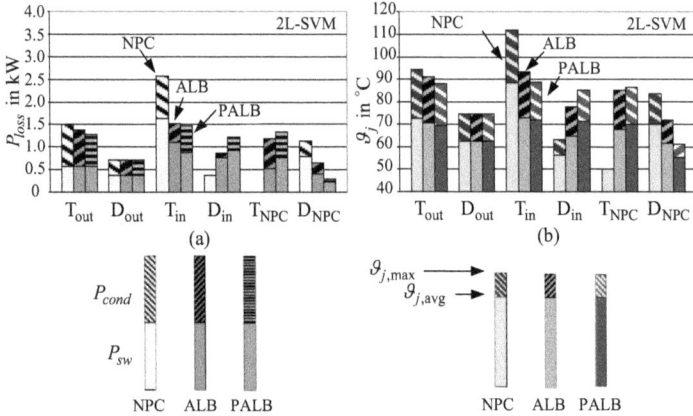

Figure 5.20 (a) Loss and (b) junction temperature distribution in a 3.3 kV 3L-ANPC VSC using 4.5 kV PP IGBTs; U_{dc}=5000 V, $i_{ph,rms}$=1200 A, f_1=2 Hz, f_{sw}=450 Hz, $\vartheta_a = 50$ °C, pf=1, m_a=0.05, 2L-SVM

Figure 5.21 shows the junction temperature ripples at low output frequency. The maximum junction temperature ripple is reduced from 36.7 °C to 29.5 °C by 19.4% applying the ALB method and from 36.7 °C to 25.9 °C by 29.3% applying the PALB method. Thus, the PALB method has an improvement of 9.9% compared to the ALB method (see Table 5.11).

Figure 5.22 shows the junction temperatures of T_2 applying the ALB and the PALB method, the reference waveform for the 2L-SVM, and the normalized phase current. The reference voltage for 2L-SVM alternates between the carrier bands with the frequency f_{comm}=25 Hz.

106

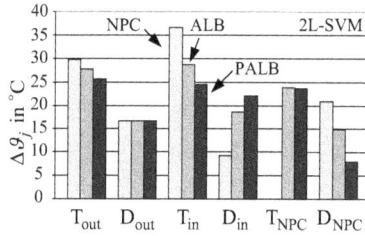

Figure 5.21 Junction temperature ripple in a 3.3 kV 3L-ANPC VSC using 4.5 kV PP IGBTs; U_{dc}=5000 V, $i_{ph,rms}$=1200 A, f_1=2 Hz, f_{sw}=450 Hz, ϑ_a = 50 °C, pf=1, m_a=0.05, 2L-SVM

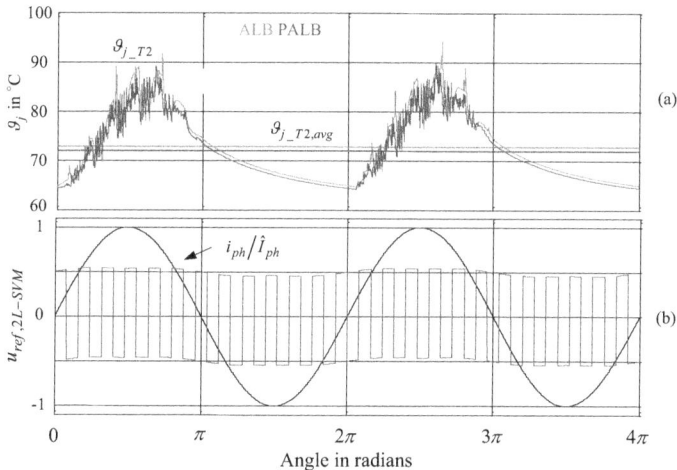

Figure 5.22 Junction temperature waveforms in a 3.3 kV 3L-ANPC VSC using 4.5 kV PP IGBTs, (a) Junction temperature of T_2; (b) Reference waveform for 2L-SVM (f_{comm}=25 Hz) and normalized phase current; U_{dc}=5000 V, $i_{ph,rms}$=1200 A, f_1=2 Hz, f_{sw}=450 Hz, ϑ_a = 50 °C, pf=1, m_a=0.05, 2L-SVM

Figure 5.23 shows the commutation types used by the balancing methods. In **Sector 1 and 4**, the PALB method uses a combination of 63% of Type 2 commutations, 23% of Type 3 commutations, and 14% of Type 1 commutations. The goal of this combination is to reduce the

107

maximal junction temperatures of T_{in}, D_{in}, and T_{out}. On the other hand, the high utilization percentage of Type 2 commutation results in a junction temperature imbalance between D_{NPC} and D_{in}. The ALB method uses mainly Type 1 and Type 3 (and only 7% Type 2) commutations to balance the junction temperature between T_{in} and T_{out}, and between D_{in} and D_{NPC}.

At **Sector 2 and 3**, both balancing methods use a combination of Type 1 and Type 2 commutations to achieve a thermal equilibrium between T_{in} and T_{NPC}. The difference is that the PALB method uses 40% percentage of Type 2 commutation. The PALB method attempts to reduce the maximal junction temperature of T_{NPC} and T_{in} by selecting the Type 2 and the Type 1 commutation. Summarizing, at low frequencies, the PALB method presents an improved performance compared to the ALB method (see Table 5.11).

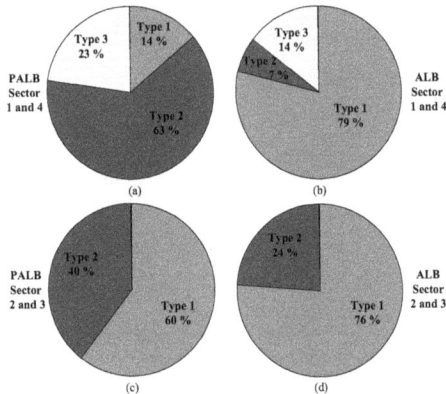

Figure 5.23 Commutations types in a 3.3 kV 3L-ANPC VSC using 4.5 kV PP IGBTs applying (a) PALB method in Sector 1 and 4; (b) ALB method in Sector 1 and 4; (c) PALB method in Sector 2 and 3; (d) ALB method in Sector 2 and 3; U_{dc}=5000 V, $i_{ph,rms}$=1200 A, f_1=2 Hz, f_{sw}=450 Hz, $\vartheta_a = 50\ ^{\circ}C$, pf=1, m_a=0.05, 2L-SVM

Table 5.11 summarizes the decrease of the maximal and average junction temperature, and the junction temperature ripple using the ALB and the PALB method.

108

Table 5.11 Junction temperature decrease applying the PALB and the ALB methods in a 3.3 kV 3L-ANPC VSC using 4.5 kV PP IGBTs

Decrease of the maximal junction temperature $\vartheta_{\max}$		
$p_{max\,ALB}$ in %	$p_{max\,PALB}$ in %	$p_{max\,PALB}$- $p_{max\,ALB}$ in %
15.8	19.5	3.7
Decrease of the average junction temperature ϑ_{avg}		
$p_{avg\,ALB}$ in %	$p_{avg\,PALB}$ in %	$p_{avg\,PALB}$- $p_{avg\,ALB}$ in %
17.7	18.7	1
Decrease of the junction temperature ripple $\Delta\vartheta_j$		
$p_{\Delta\,ALB}$ in %	$p_{\Delta\,PALB}$ in %	$p_{\Delta\,PALB}$ - $p_{\Delta\,ALB}$ in %
19.4	29.3	9.9

5.3.2.2 Evaluation of a 3.3 kV 3L-ANPC VSC using 4.5 kV IGBT modules

The performance of a 3.3 kV 3L-ANPC VSC using 4.5 kV IGBT modules is evaluated at low output frequency. The applied balancing methods are investigated in comparison. Table 5.12 summarizes the simulations parameters.

Table 5.12 Simulation parameters for a 3.3 kV 3L-ANPC VSC using 4.5 kV IGBT modules

Operating points				Simulation parameters of a 3.3 kV 3L-ANPC VSC 4.5 kV Mitsubishi CM900HB90H IGBT module				
pf	m_a	PWM scheme	f_{sw} in Hz	$i_{ph,rms}$ in A	f_l in Hz	U_{dc} in V	ϑ_a in °C	$\vartheta_{j,\max}$ in °C
±1	0.05	2L-SVM	450	625	2	4854	37	67

5. PALB method for 3L-ANPC VSC

At low output frequency, the NPC mode (see Figure 5.24) is characterized by the maximal junction temperature of T_{in}. Applying the balancing methods, the maximal junction temperature is reached in D_{in}. Both balancing methods achieve thermal equilibrium between T_{in} and T_{out}, and between D_{NPC} and D_{in}. The maximal junction temperature is reduced from 84.1 °C to 75.3 °C by 10.5% applying the ALB method and from 84.1 °C to 74.4 °C by 11.6% applying the PALB method (see Table 5.13). Thus, the PALB method has an improvement of 1.1% compared to the ALB method.

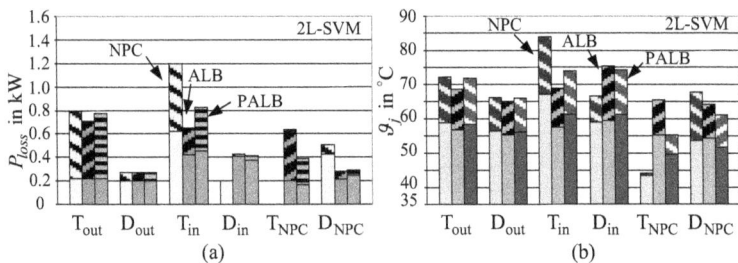

Figure 5.24 (a) Loss and (b) junction temperature distribution a 3.3 kV 3L-ANPC VSC using 4.5 kV IGBT modules; U_{dc}=4854 V, $i_{ph,rms}$=625 A, f_1=2 Hz, f_{sw}=450 Hz, $\vartheta_a = 37°C$, pf=1, m_a=0.05, 2L-SVM

Figure 5.25 shows the junction temperature ripple at low output frequency. The maximum junction temperature ripple is reduced from 26.5 K to 22 K by 16.8% applying the ALB method and from 26.5 K to 19.7 K by 25.5% applying the PALB method. Thus, the PALB method has an improvement of 8.7% compared to the ALB method (see Table 5.13).

Figure 5.26 shows the junction temperatures of the critical device D_2 applying the ALB and the PALB method, the reference waveform of the 2L-SVM, and the normalized phase current. The reference voltage for 2L-SVM alternates between the carrier bands with a frequency of f_{comm}=25 Hz.

Figure 5.27 shows the commutation types used by the balancing methods in case C. In **Sector 1 and 4**, the PALB method uses a

110

combination of Type 1 and Type 2 commutations to decrease the junction temperatures of D_{in} and T_{in}, respectively. However, during the Type 2 commutation, D_{in} is stressed with conduction and switching losses. Since D_{in} is thermally coupled with T_{in}, the accumulated losses in the diode will increase in time the junction temperature of T_{in}. On the other hand, the ALB method uses a combination of Type 1 and Type 3 commutations to achieve thermal balance between T_{in} and T_{out}, and between D_{in} and D_{NPC}.

In **Sector 2 and 3**, both balancing methods use a combination of Type 1 and Type 2 commutations to achieve thermal equilibrium between T_{in} and T_{NPC}, and between D_{in} and D_{out}. However, the PALB method presents a larger percentage of Type 2 commutations to reduce the junction temperature of D_{in}. However, during the Type 2 commutation, T_{in} is stressed with conduction and switching losses. Since T_{in} is thermally coupled with D_{in}, the accumulated losses in the power transistor will increase the junction temperature of D_{in}. As a consequence, T_{in} and D_{in} present a lower maximal junction temperature, but a higher average junction temperature applying the PALB method compared to ALB method.

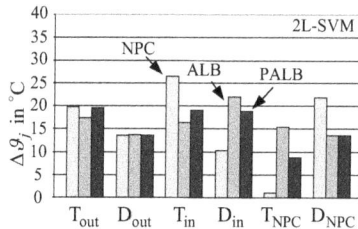

Figure 5.25 Junction temperature ripple in a 3.3 kV 3L-ANPC VSC using 4.5 kV IGBT modules; U_{dc}=4854 V, $i_{ph,rms}$=625 A, f_l=2 Hz, f_{sw}=450 Hz, $\vartheta_a = 37°C$, pf=1, m_a=0.05, 2L-SVM

Table 5.13 summarizes the decrease of the maximal and average junction temperature, and the junction temperature ripple applying the balancing methods. The PALB method presents an improved performance compared to the ALB method regarding the maximal junction temperature decrease.

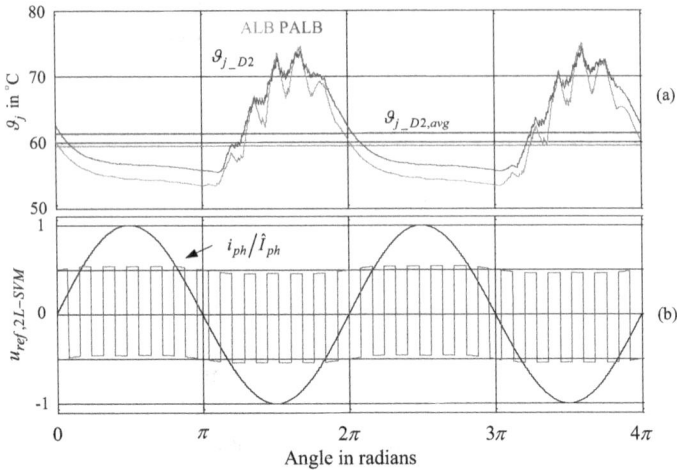

Figure 5.26 Junction temperature waveforms in a 3.3 kV 3L-ANPC VSC using 4.5 kV IGBT modules, (a) Junction temperature of D2; (b) Normalized phase midpoint output voltage and normalized phase current; U_{dc}=4854 V, $i_{ph,rms}$=625 A, f_1=2 Hz, f_{sw}=450 Hz, $\vartheta_a = 37°C$, pf=1, m_a=0.05, 2L-SVM

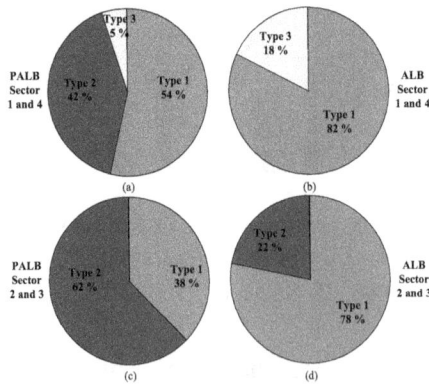

Figure 5.27 Commutations types in a 3.3 kV 3L-ANPC VSC using 4.5 kV IGBT modules applying (a) PALB method in Sector 1 and 4; (b) ALB method in Sector 1 and 4; (c) PALB method in Sector 2 and 3; (d) ALB method Sector 2 and 3; U_{dc}=4854 V, $i_{ph,rms}$=625 A, f_1=2 Hz, f_{sw}=450 Hz, $\vartheta_a = 37°C$, pf=1, m_a=0.05, 2L-SVM

112

Table 5.13 Junction temperature decrease applying the PALB and the ALB methods in a 3.3 kV 3L-ANPC VSC using 4.5 kV IGBT modules

Decrease of the maximal junction temperature ϑ_{max}		
$p_{max\ ALB}$ in %	$p_{max\ PALB}$ in %	$p_{max\ PALB}$-$p_{max\ ALB}$ in %
10.5	11.6	1.1
Decrease of the average junction temperature ϑ_{avg}		
$p_{avg\ ALB}$ in %	$p_{avg\ PALB}$ in %	$p_{avg PALB}$-$p_{avg\ ALB}$ in %
11.4	8.5	-2.9
Decrease of the junction temperature ripple $\Delta\vartheta_j$		
$p_{\Delta\ ALB}$ in %	$p_{\Delta\ PALB}$ in %	$p_{\Delta\ PALB}$ - $p_{\Delta\ ALB}$ in %
16.8	25.5	8.7

5.3.3 Operation at zero speed

5.3.3.1 Evaluation of a 3.3 kV 3L-ANPC VSC using 4.5 kV press-pack IGBTs

The performance of a 3.3 kV 3L-ANPC VSC using 4.5 kV PP IGBTs is evaluated at zero speed. The results of the PALB method are analyzed in comparison with the ALB method. Table 5.14 summarizes the simulations parameters.

Figure 5.28 shows the loss and junction temperature distribution of a 3.3 kV 3L-ANPC VSC. At zero speed and positive phase current, the inner switch T_2 is the most stressed device in all operation modes. The PALB and the ALB methods achieve thermal balance between T_1, T_2, and T_6. Thus, the maximal junction temperature is reduced from 124.5 °C to 98.7 °C by 20.7% applying the ALB method and from 124.5 °C to 93.2 °C by 25.1% applying the PALB method. Thus, the PALB method presents an improvement of 4.4% compared to the ALB method (see Table 5.15)

Table 5.14 Simulation parameters for a 3.3 kV 3L-ANPC VSC using 4.5 kV PP IGBTs

Operating points				Simulation parameters of a 3.3 kV 3L-NPC VSC 4.5 kV PP Westcode T1200EB45E IGBTs and 4.5 kV PP Infineon D1031SH45T diodes				
pf	m_a	PWM scheme	f_{sw} in Hz	$i_{ph,rms}$ in A	f_1 in Hz	U_{dc} in V	ϑ_a in °C	$\vartheta_{j,\max}$ in °C
±1	0.05	2L-SVM	450	760	0	5000	50	125

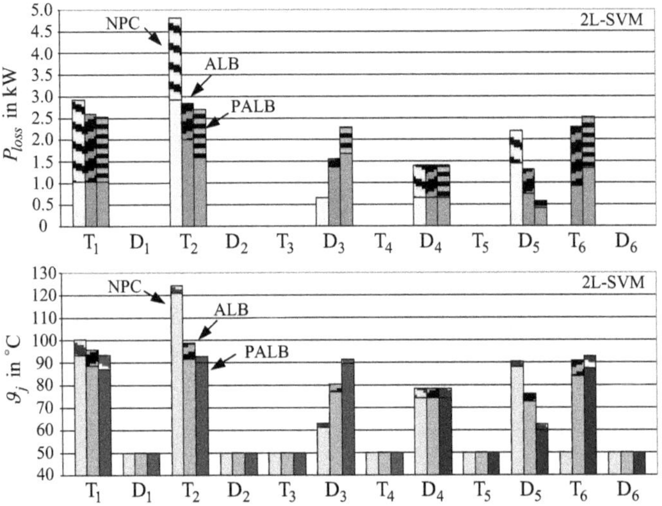

Figure 5.28 Loss and junction temperature distribution in a 3.3 kV 3L-ANPC VSC using 4.5 kV PP IGBTs; U_{dc}=5000 V, $i_{ph,rms}$=1200 A, f_1=0 Hz, f_{sw}=450 Hz, $\vartheta_a = 50\ ^\circ\mathrm{C}$, *pf*=1, m_a=0.05, 2L-SVM

Figure 5.29 pictures the critical junction temperatures of T_2 applying the ALB and the PALB methods, the reference waveform for 2L-SVM, and the normalized phase current. The phase current is considered to remain constant for the entire output period.

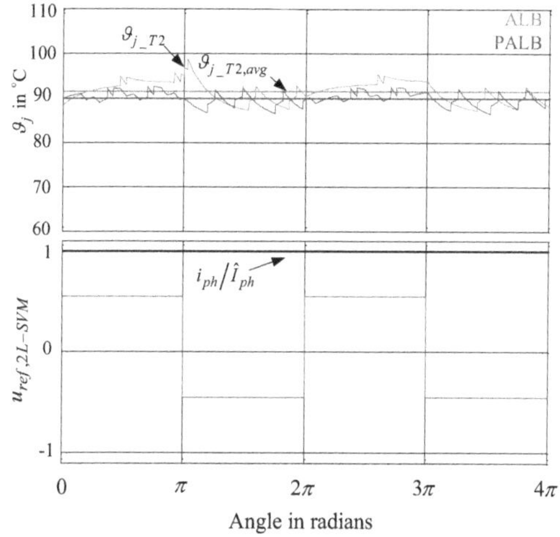

Figure 5.29 Junction temperature waveforms in a 3.3 kV 3L-ANPC VSC

Figure 5.29 Junction temperature waveforms in a 3.3 kV 3L-ANPC VSC using 4.5 kV PP IGBTs, (a) Junction temperature of T_2; (b) Reference waveform for 2L-SVM (f_{comm}=25 Hz) and normalized phase current; U_{dc}=5000 V, $i_{ph,rms}$=1200 A, f_1=0 Hz, f_{sw}=450 Hz, ϑ_a = 50 °C, pf=1, m_a=0.05, 2L-SVM

Figure 5.30 presents the percentage of commutation types applying the PALB and the ALB method, respectively. In **Sector 1**, the PALB method uses a combination of 64% of Type 2, 30% of Type 1, and 14% of Type 3 commutations to reduce the maximal junction temperatures of T_2, D_3 and T_1, respectively. On the other hand, the increased percentage of Type 2 commutation drives into imbalance the junction temperature of D_5 and D_3. The ALB method uses a combination of Type 1 and Type 3 commutations to balance the junction temperature between T_1 and T_2, and between D_3 and D_5.

In **Sector 3**, both methods use a combination of Type 1 and Type 2 commutations to achieve thermal balance between T_2 and T_6. The difference is that the PALB method uses 44.6% percentage of Type 2 commutation to reduce the maximal junction temperature of T_6. The increased junction temperature of T_6 is a result of the selection of the Type

115

5. PALB method for 3L-ANPC VSC

2 commutation in Sector 1, during which T_6 has been stressed with conduction losses.

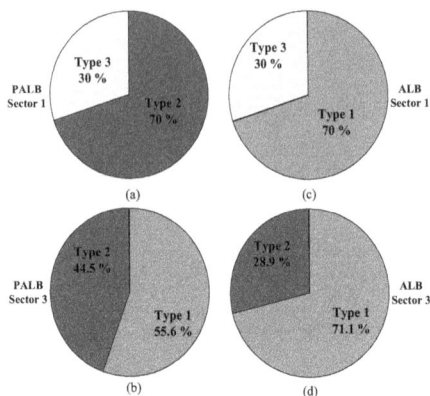

Figure 5.30 Commutations types in a 3.3 kV 3L-ANPC VSC using 4.5 kV IGBT modules applying (a) PALB method in Sector 1; (b) ALB method in Sector 1; (c) PALB method in Sector 3; (d) ALB method in Sector 3; U_{dc}=5000 V, $i_{ph,rms}$=1200 A, f_1=0 Hz, f_{sw}=450 Hz, $\vartheta_a = 50\ ^\circ C$, pf=1, m_a=0.05, 2L-SVM

As a conclusion, the PALB method presents at zero speed an improved performance compared to the ALB method (see Table 5.15).

Table 5.15 Junction temperature decrease applying the PALB and the ALB methods in a 3.3 kV 3L-ANPC VSC using 4.5 PP IGBTs

Decrease of the maximal junction temperature ϑ_{max}		
$p_{max\ ALB}$ in %	$p_{max\ PALB}$ in %	$p_{max\ PALB}$- $p_{max\ ALB}$ in %
20.7	25.1	4.4
Decrease of the average junction temperature ϑ_{avg}		
$p_{avg\ ALB}$ in %	$p_{avg\ PALB}$ in %	$p_{avg\ PALB}$- $p_{avg\ ALB}$ in %
24.4	25.8	1.4

116

5.3.3.2 Evaluation of a 3.3 kV 3L-ANPC VSC using 4.5 kV IGBT modules

The performances of the balancing methods are evaluated for a 3.3 kV 3L-ANPC VSC using 4.5 kV IGBT modules at zero speed. Table 5.16 summarizes the simulations parameters.

Table 5.16 Simulation parameters for a 3.3 kV 3L-ANPC VSC using 4.5 kV IGBT modules

Operating points								
				Simulation parameters of a 3.3 kV 3L-ANPC VSC 4.5 kV Mitsubishi CM900HB90H IGBT module				
pf	m_a	PWM scheme	f_{sw} in Hz	$i_{ph,rms}$ in A	f_l in Hz	U_{dc} in V	ϑ_a in °C	$\vartheta_{j,max}$ in °C
±1	0.05	2L-SVM	450	230	0	4854	37	67

Figure 5.31 illustrates the loss and junction temperature distribution in a 3.3 kV 3L-ANPC VSC. At zero speed and positive phase current, T_2 is the most stressed device in all operation modes. The PALB and the ALB methods achieve thermal balance between T_1, T_2, and T_6. The maximal junction temperature is decreased from 68.2 °C to 56.4 °C by 17.3% applying the ALB method and from 68.2 °C to 55.4 °C by 18.7% applying the PALB method. Thus, the PALB method presents an improvement of 1.4% compared to the ALB method (see Table 5.17).

Figure 5.32 pictures the critical junction temperatures of T_2 applying the ALB and the PALB method, the corresponding reference waveform for 2L-SVM, and the normalized phase current. The phase current is considered to remain constant during the entire output period.
In **Sector 1**, the ALB method uses a combination of Type 1 and Type 3 commutations to achieve thermal balance between T_1 and T_2, and between D_3 and D_5. The PALB method chooses the Type 2 instead of the Type 1 commutations in order to decrease the highest junction temperature of T_2. In **Sector 3**, both methods use a combination of Type

117

5. PALB method for 3L-ANPC VSC

1 and Type 2 commutations to achieve thermal balance between T_2 and T_6. The PALB method selects the Type 2 instead of the Type 1 commutation in order to reduce the junction temperature of T_6. In conclusion, the PALB method presents at zero speed an improved performance compared to the ALB method.

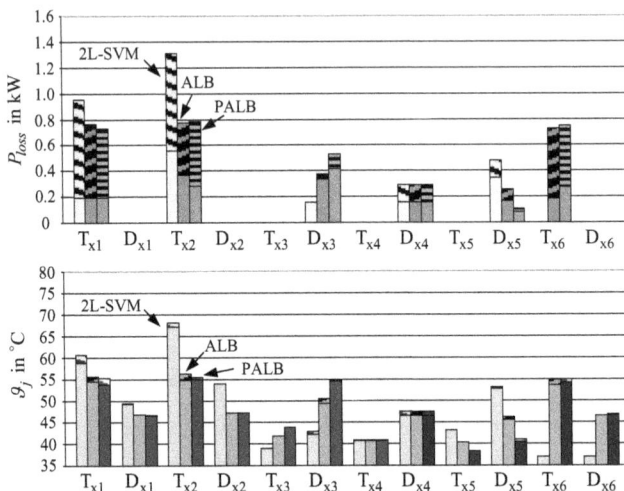

Figure 5.31 Loss and junction temperature distribution in a 3.3 kV 3L-ANPC VSC using 4.5 kV IGBT modules; U_{dc}=4854 V, $i_{ph,rms}$=230 A, f_1=0 Hz, f_{sw}=450 Hz, $\vartheta_a = 37°C$, pf=1, m_a=0.05, 2L-SVM

Table 5.17 summarizes the decrease of the maximal and average junction temperature applying the balancing methods.

Table 5.17 Junction temperature decrease applying the PALB and the ALB methods in a 3.3 kV 3L-ANPC VSC using 4.5 kV IGBT modules

Decrease of the maximal junction temperature ϑ_{max}		
$p_{max\ ALB}$ in %	$p_{max\ PALB}$ in %	$p_{max\ PALB} - p_{max\ ALB}$ in %
17.3	18.7	1.4
Decrease of the average junction temperature ϑ_{avg}		
$p_{avg\ ALB}$ in %	$p_{avg\ PALB}$ in %	$p_{avg\ PALB} - p_{avg\ ALB}$ in %
18.2	18.2	0

118

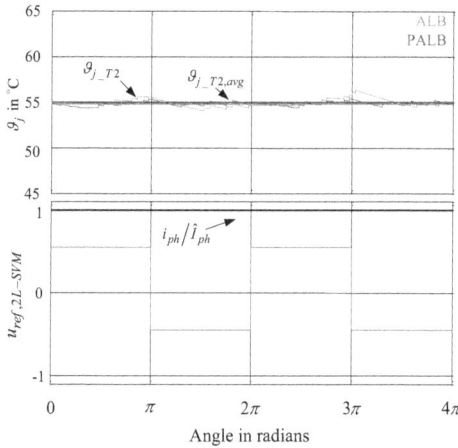

Figure 5.32 Junction temperature waveforms in a 3.3 kV 3L-ANPC VSC using 4.5 kV IGBT modules, (a) Junction temperature of T_2; (b) Normalized phase midpoint output voltage and normalized phase current; U_{dc}=4854 V, $i_{ph,rms}$=230 A, f_1=0 Hz, f_{sw}=450 Hz, $\vartheta_a = 37°C$, pf=1, m_a=0.05, 2L-SVM

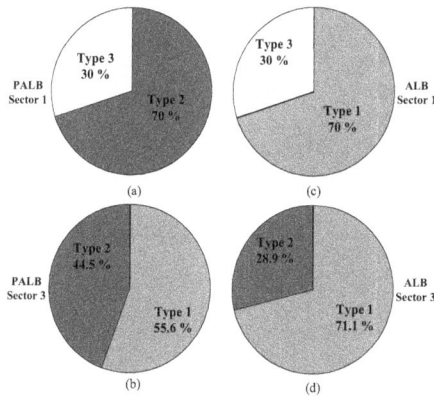

Figure 5.33 Commutations types in a 3.3 kV 3L-ANPC VSC using 4.5 kV IGBT modules applying (a) PALB method in Sector 1; (b) ALB method in Sector 1; (c) PALB method in Sector 3; (d) ALB method in Sector 3; U_{dc}=4854 V, $i_{ph,rms}$=230 A, f_1=0 Hz, f_{sw}=450 Hz, $\vartheta_a = 37°C$, pf=1, m_a=0.05, 2L-SVM

119

5.4 Converter Output Power

The ALB and the PALB methods are compared regarding the maximum achievable phase current and output power. The phase current is limited by the maximum allowable junction temperature $\vartheta_{j,max}$ and the maximum rise of the junction temperature:

$$\Delta \vartheta_{j,max} = \vartheta_{j,avg} - \vartheta_a \tag{5.21}$$

where ϑ_{avg} is the maximum allowable average junction temperature and ϑ_a is the ambient (water) temperature.

In order to evaluate the converter's performance applying the balancing methods with respect to the NPC operation mode, the output power gain is introduced. The power gain is defined as the ratio between the maximum output power of the NPC operation mode and the maximum output power in the ALB or the PALB operation mode. The output power gain is expressed in percentage.

$$gain_{power\_ALB} = \frac{S_{c(ALB)}}{S_{c(NPC)}}(\%), \tag{5.22}$$

$$gain_{power\_PALB} = \frac{S_{c(PALB)}}{S_{c(NPC)}}(\%), \tag{5.23}$$

where $S_{c(NPC)}$, $S_{c(ALB)}$, and $S_{c(PALB)}$ are the output power of the 3L-ANPC VSC using the NPC, the ALB, and the PALB operation mode, respectively.

5.4.1 Evaluation of a 3.3 kV 3L-ANPC VSC using 4.5 kV press-pack IGBTs

It is assumed that the maximum temperature difference between junction and ambient for the PP IGBT should not exceed 40 K [91]. Thus, the maximum allowable average junction temperature is 90 °C for an ambient temperature of 50 °C. Due to this thermal restriction, the phase current has to be limited according to the operation point, resulting in power derating. At maximum modulation index, the largest thermal imbalance occurs in case A, limiting the phase current. Table 5.18 compares the maximum allowable phase current, output power, and power ratio of a 3.3 kV 3L-ANPC VSC using PP IGBTs. The switching frequency range is between 300 Hz up to 1050 Hz. The rated PP IGBT current is $I_{C,n}$=1200 A.

To begin with, the performance of the PALB method is evaluated at maximum modulation index m_a=1.15. At f_{sw}=450 Hz (3L-SVM), the PALB method enables a rise of the phase current from 930 A to 1140 A, and thus, a 23 % higher output power. Furthermore, at f_{sw}=500 Hz (SHE-5α) and f_{sw}=750 Hz (3L-SVM), the phase current can be increased up to 1170 A by 23% and to 980 A by 40 %, respectively. At f_{sw}=1050 Hz, even applying the balancing method, the phase current is limited to 840 A. The PALB method features an improvement up to 2 % compared to the ALB method.

Table 5.18 Maximum output power of a 3.3 kV 3L-ANPC VSC using 4.5 kV PP IGBTs

Simulation parameters of a 3.3 kV 3L-ANPC VSC using 4.5 kV PP Westcode T1200EB45E IGBTs and 4.5 kV PP Infineon D1031SH45T diodes U_{dc}=5000 V, f_f=50 Hz, $\vartheta_a = 50°C$, $\vartheta_{j,\text{avg}} = 90°C$										
Operating points			NPC		ALB			PALB		
p.f	m_a	PWM scheme	$i_{ph,rms}$ in A	S_c in MVA	$i_{ph,rms}$ in A	S_c in MVA	$\frac{S_{c(ALB)}}{S_{c(NPC)}}$ in %	$i_{ph,rms}$ in A	S in MVA	$\frac{S_{c(ALB)}}{S_{c(NPC)}}$ in %
f_{sw} =300 Hz										
±1	0.05	2L-SVM	1200	6.86	1200	6.86	100	1200	6.86	100
1	1.15	SHE-3α	1130	6.46	1200	6.86	106	1200	6.86	106
f_{sw} =450 Hz										
±1	0.05	2L-SVM	1200	6.86	1200	6.86	100	1200	6.86	100
1	1.15	3L-SVM	930	5.32	1130	6.46	122	1140	6.52	123

121

5. PALB method for 3L-ANPC VSC

			f_{sw} =500 Hz							
±1	0.05	2L-SVM	1200	6.86	1200	6.86	100	1200	6.86	100
1	1.15	SHE-5α	950	5.43	1160	6.63	122	1170	6.68	123
			f_{sw} =750 Hz							
±1	0.05	2L-SVM	1050	6.00	1200	6.86	114	1200	6.86	114
1	1.15	3L-SVM	700	4.00	970	5.54	139	980	5.60	140
			f_{sw} =1050 Hz							
±1	0.05	2L-SVM	900	5.14	1200	6.86	133	1200	6.86	133
1	1.15	3L-SVM	550	3.14	830	4.74	151	840	4.80	153
			Zero speed: f_1 =0 Hz, f_{sw} =450 Hz							
±1	0.05	2L-SVM	470	2.69	740	4.23	157	760	4.34	162

At small modulation index m_a=0.05, the balancing methods present similar performances and accomplish an increase of the output power up to 33 %. The phase current is set to the nominal current $I_{ph,rms}$=1200 A for the entire range of switching frequency up to 1050 Hz.

At zero speed, both balancing methods achieve noticeable improvements. The PALB method enables an increase of the phase current from 470 A to 760 A and thus, a 57 % higher output power. The PALB method presents an improved performance of 5 % compared to the ALB method.

5.4.2 Evaluation of a 3.3 kV 3L-ANPC VSC using 4.5 kV IGBT modules

The maximum temperature difference between junction and ambient for the IGBT module should not exceed 30 K due to thermal stress [96]. Thus, the maximum allowable average junction temperature is calculated to 67 °C for an ambient water temperature of 37 °C. This thermal restriction limits the phase current according to the operation point. At maximum modulation index m_a=1.15, the limiting operating point is case A (pf=1) for 3L-SVM and case B (pf=−1) for SHE. In case

A, the outer switches reach their maximal junction temperature, whereas in case B, the inner diodes are the most stressed power devices.

Table 5.19 Maximum output power in a 3.3 kV 3L-ANPC VSC using 4.5 kV IGBT module

			NPC		ALB			PALB		
		Simulation parameters of a 3.3 kV 3L-ANPC VSC 4.5 kV Mitsubishi CM900HB90H IGBT module U_{dc}=5000 V, f_1=50 Hz, $\vartheta_a = 37°C$, $\vartheta_{j,\text{avg}} = 67°C$								
Operating points			NPC		ALB			PALB		
pf	m_a	PWM scheme	$i_{ph,rms}$ in A	S_c in MVA	$i_{ph,rms}$ in A	S_c in MVA	$\frac{S_{c(ALB)}}{S_{c(NPC)}}$ in %	$i_{ph,rms}$ in A	S in MVA	$\frac{S_{c(ALB)}}{S_{c(NPC)}}$ in %
			f_{sw} =300 Hz							
±1	0.05	2L-SVM	710	4.06	820	4.69	114	710	4.06	100
−1	1.15	SHE-3α	560	3.20	630	3.60	113	580	3.32	104
			f_{sw} =450 Hz							
±1	0.05	2L-SVM	625	3.58	800	4.57	128	740	4.22	119
1	1.15	3L-SVM	460	2.63	620	3.54	135	610	3.49	133
			f_{sw} =500 Hz							
±1	0.05	2L-SVM	600	3.43	770	4.40	128	720	4.12	120
−1	1.15	SHE-5α	480	2.74	600	3.43	125	550	3.14	115
			f_{sw} =750 Hz							
±1	0.05	2L-SVM	500	2.86	690	3.89	136	660	3.77	132
1	1.15	3L-SVM	310	1.77	480	2.74	155	480	2.74	155
			f_{sw} =1050 Hz							
±1	0.05	2L-SVM	400	2.29	560	3.20	140	560	3.20	140
1	1.15	3L-SVM	220	1.26	390	2.23	177	390	2.23	177
			Zero speed: f_1 =0 Hz, f_{sw} =450 Hz							
±1	0.05	2L-SVM	230	1.31	390	2.23	169	370	2.11	161

5. PALB method for 3L-ANPC VSC

Table 5.19 summarizes the maximum allowable phase current, output power and power ratio of a 3.3 kV 3L-ANPC VSC. The nominal IGBT module current is $I_{C,n}$=900 A. The output power applying the PALB method and SVM at f_{sw}=450 Hz is enlarged to 133%. At f_{sw}=750 Hz, the balancing methods present similar results: the phase current is increased from 310 A to 480 A by 55 %. Increasing the switching frequency to 1050 Hz, even that both balancing methods achieve an output power gain of 77%, the maximal phase current is limited to 390 A. Despite the improved performance obtained by PALB method compared to the conventional NPC, the ALB method achieves an overall better result.

5.5 Conclusions

The structure and the performance of the new proposed balancing method are presented. The efficiency of the balancing methods is evaluated concerning the junction temperature and the maximum achievable phase current. The 3L-ANPC VSC is investigated at grid frequency, at low output frequency, and at zero speed. Moreover, the thermal behavior of the 3L-ANPC VSC is analyzed for two packaging types: module and press-pack, due to the different thermal coupling and thermal restrictions.

The 3.3 kV 3L-ANPC VSC featuring 4.5 kV press-pack IGBTs and applying the PALB method presents a decrease of the maximal junction temperature by up to 20.2% compared to the 3L-NPC VSC and by up to 3.6% compared to the use of the ALB method. Furthermore, at low output frequency the PALB method has an improvement of 29.3% reducing the junction temperature ripple compared to conventional NPC and 9.9% better than the ALB method. At zero speed, the PALB method reduces by 4.4% the maximal junction temperature. The decrease of the junction temperature enables an output power increase: thus, the PALB method achieves an output power increase by up to 53% compared to conventional NPC and by up to 5% compared to ALB method.

Although the PALB has an overall better performance compared to the 3L-NPC mode, for the 3.3 kV 3L-ANPC VSC featuring 4.5 kV IGBT modules operating in rectifier mode, the ALB decreases the maximal junction temperature by up to 9.2% compared to the PALB method. Thus, the ALB achieves a power increase up to 14% compared to the PALB method.

Depending on the operation point and thermal characteristics of the employed power semiconductors, one method presents slight advantages over the other. Summarizing, the 3 kV 3L-ANPC VSC featuring 4.5 kV press-pack IGBTs using the PALB method achieves the most important improvements at low output frequency and zero speed. The improvements are mainly due to use of a higher percentage of the Type 2 commutation. However, the PALB method presents a high computational time since the method predicts the thermal behavior of the converter, making it more difficult to implement it on a microcontroller.

One possibility of a practical implementation is the use of a look-up table. The PALB method is compiled off-line to achieve the optimal utilization percentage or the sequence of the commutations types. This information can be stored in a look-up table for different operation points as a function of m_a and pf. The optimal commutation sequence is on-line selected depending on the operation point. The main drawback of the look-up table is that it works only for pre-calculated operation points. However, the processing time can be significantly reduced compared to the runtime of the predictive algorithm.

6 Conclusions

This thesis presents the evaluation of two 3.3 kV 3L-ANPC VSCs featuring 4.5 kV press-pack IGBTs and 4.5 kV IGBT modules, respectively. Compared to the conventional 3L-NPC converter, the 3L-ANPC converter presents additional switch states and commutations that allow a suitable distribution of the losses among the semiconductors. The loss and junction temperature distribution can be controlled by the selection of the zero state using a loss balancing method. Thus, the balancing method represents a practical measure to optimize the performance of the 3L-ANPC, allowing an increase of the converter output power and alternatively, an increase of the switching frequency.

There are two different loss balancing methods evaluated in this work: the Active Loss Balancing (ALB) method [60] and a new method, the predictive active balancing (PALB) method. The control principle of the ALB method is based on a decision chart that distributes the switching losses among the semiconductors to balance the junction temperatures. Thus, a significant reduction of the highest junction temperature is achieved. However, the ALB method does not consider the conduction losses explicitly and could fail to relief the optimal zero states at low output frequencies and zero speed operation. Instead, the PALB method considers for the selection of the zero state both the conduction and the switching losses. The PALB method predicts the power losses during the time interval up to the next commutation to a zero state and based on the Foster thermal model, predicts the junction temperatures. Since each of the new available zero states delivers a different junction temperature distribution, the PALB method selects as optimum the zero states which leads to the lowest junction temperature of the power semiconductor of one phase leg.

In the first comparison, the efficacy of the balancing methods is evaluated regarding to the junction temperature. At grid frequency (f_1=50 Hz) and maximal modulation index (m_a=1.15), in the case of the 3.3 kV 3L-ANPC featuring press-pack IGBTs, the use of the PALB method

126

allows a reduction of the highest junction temperature by 0.5% compared to the ALB method. Further improvements are achieved at minimum modulation index (m_a=0.05) by applying the PALB method, resulting in a lower junction temperature from 1.2% to 3.6% compared to the ALB method. Therefore, in the case of the converter using press-pack IGBTs, the PALB method delivers slightly better results compared to the ALB method.

For the 3L-ANPC VSC featuring IGBT modules operating at grid frequency, the ALB method presents in general a better performance compared to the PALB method. At maximum modulation index, the ALB method allows a junction temperature reduction from 0.1% to 9.2% compared to the PALB method. Furthermore, at minimum modulation index, the ALB method decreases the highest junction temperature by 2.3%.

The PALB method presents the most important improvements at low output frequency (f_1=2 Hz, m_a=0.05, f_{sw}=450 Hz, 2L-SVM) and at zero speed (f_1=0 Hz, m_a=0.05, f_{sw}=450 Hz, 2L-SVM). During the operation at low frequency, the 3L-ANPC VSC featuring press-pack IGBTs presents a reduction of the highest junction temperature by 3.7% using the PALB method instead of the ALB method. Moreover, the PALB method has an improvement of 9.9% regarding the decrease of the junction temperature ripple. At zero speed, the highest junction temperature is reduced by 4.4% applying the PALB method. In the case of the 3L-ANPC converter featuring IGBT module, investigations showed that the PALB method decreases the highest junction temperature by 1.1% and the junction temperature ripple by 8.7% compared to the ALB method. At zero speed, the 3L-ANPC converter presents a 1.4% lower junction temperature applying the PALB method compared to the ALB method.

In the second comparison, the performances of the balancing methods are evaluated regarding to the maximum allowable phase current considering the thermal constrains. It is assumed that the junction temperature difference between junction and ambient (water) of the press-pack IGBT should not exceed 40 K. The investigations showed that at maximum modulation index the phase current can be increased by 1%

6. Conclusions

using of the PALB method instead of the ALB method. As an example, the maximum phase current of the 3L-NPC VSC is restricted to 950 A at f_{sw}=500 Hz (SHE-5α); applying the PALB method and the ALB method, the phase current can be increased to 1170 A and 1160 A, respectively. Next, at minimum modulation index, the balancing methods present similar performances: the phase current is set to the nominal current $i_{ph,rms}$=1200 A. At zero speed, the maximum phase current of the 3L-NPC VSC is limited to 470 A. The PALB method enables an increase to 760 A and thus, a 57% higher phase current. Moreover, the PALB method presents an improved performance of 5% compared to the ALB method (740 A).

In the case of the 3L-ANPC VSC featuring IGBT modules, the temperature difference between junction and ambient is set to 30 K. At maximum modulation index, the ALB method allows a phase current increase from 2% to 10% compared to the PALB method. Furthermore, at minimum modulation index, the ALB method achieves a phase current increase from 4% to 14%. At zero speed, the phase current is restricted to 230 A in the NPC mode. Both balancing methods achieve major improvements. The ALB method enables an increase of the phase current up to 390 A by 69 %. On the other hand, the PALB method has a power gain of 61%, i.e. 8 % less compared to the ALB method.

The use of the balancing methods applied to the 3L-ANPC VSC presents substantially better results in all critical operation points as compared to the performance of the 3L-NPC VSC. Thus, the phase current and the output power range can be increased without exceeding the thermal limitations of the semiconductors. Furthermore, the loss balancing methods reduce the junction temperature ripple and consequently, the thermo-mechanical stress of the semiconductors. Therefore, the lifetime expectancy of the semiconductors and the reliability of the converter are increased.

However, in general one balancing method has slight advantages over the other depending on the operation point and the packaging type of the power semiconductors. The most important improvements were obtained by the PALB method applied to the 3.3 kV 3L-ANPC VSC using 4.5 kV press-pack IGBTs at small modulation index, small output frequency, and zero speed in comparison to the ALB method. The reason is that the PALB method explicitly considers the conduction losses that are significant at small modulation depth. Furthermore, the evaluation of the PALB method has shown that at small output frequency, the use of a higher percentage of Type 2 commutation has substantially reduced the junction temperature ripple. The main drawback of the proposed predictive method remains the higher computational time making the implementation in a microcontroller a complex task. A practical solution is to calculate the PALB method for different operation points and to store the optimal zero state sequences in a look-up table. This can be on-line processed and thus, the runtime can be significantly reduced.

Appendix

A Thermal behavior of a 3.3 kV 3L-ANPC VSC using 4.5 kV press-pack IGBTs

The performance of a 3.3 kV 3L-ANPC VSC using 4.5 kV press-pack IGBTs is further evaluated at grid frequency applying the PALB method and the ALB method, respectively. Tables A.1, A.4, and A.7 summarize the simulations parameters.

Tables A.2, A.3, A.5, A.6, A.8, and A.9 summarize the decrease of the maximal and average junction temperature applying the balancing methods in comparison to the conventional NPC converter. The PALB method presents generally a better performance than the ALB method. However, in case B at f_{sw}=750 Hz (see Table A.5), the ALB method decreases the maximal junction temperature by 0.3% compared to the PALB method. Thus, the use of exclusively Type 1 and Type 3 commutations presents better results than the combination of all types of commutations.

Table A.1 Simulation parameters for a 3.3 kV 3L-ANPC VSC at 300 Hz

Operating points				Simulation parameters of a 3.3 kV 3L-NPC VSC 4.5 kV PP Westcode T1200EB45E IGBTs and 4.5 kV PP Infineon D1031SH45T diodes					
Case	pf	m_a	PWM scheme	f_{sw} in Hz	$i_{ph,rms}$ in A	f_1 in Hz	U_{dc} in V	ϑ_a in °C	$\vartheta_{j,max}$ in °C
A	+1	1.15	SHE-3α						
B	−1	1.15	SHE-3α						
C	+1	0.05	SHE-3α	300	1200	50	5000	50	125
D	−1	0.05	SHE-3α						
E	±1	0.05	2L-SVM						

Table A.2 Percentage decrease of the maximal junction temperature

	NPC		ALB			PALB			Gain in %
	Power devices	ϑ_{max} in °C	Power devices	ϑ_{max} in °C	p_{max} ALB in %	Power devices	ϑ_{max} in °C	p_{max} PALB in %	
A	T_{out}	100.7	T_{in}	94.6	6.1	T_{out}	94.2	6.5	0.4
B	D_{out}	86.5	D_{in}	83.9	3.0	D_{in}	83.6	3.4	0.4
C	D_{NPC}	91.9	T_{out}	79.6	13.5	T_{in}	77.8	15.3	1.8
D	T_{in}	118.2	T_{NPC}	86.5	26.8	T_{NPC}	86.5	26.8	0
E	T_{in}	91.7	T_{out}	77.4	15.6	T_{in}	76.9	16.9	1.3

130

Table A.3 Percentage decrease of the average junction temperature

	NPC		ALB			PALB			
	Power devices	$\vartheta_{j,avg}$ in °C	Power devices	$\vartheta_{j,avg}$ in °C	p_{max} ALB in %	Power devices	$\vartheta_{j,avg}$ in °C	p_{max} PALB in %	Gain in %
A	T_{out}	93.3	T_{in}	87.2	6.5	T_{out}	86.7	7.1	0.6
B	D_{out}	82.2	D_{in}	79.4	3.3	D_{in}	79.1	3.7	0.4
C	D_{NPC}	86.4	T_{out}	71.9	16.8	T_{in}	71.4	17.4	0.6
D	T_{in}	105.0	T_{NPC}	77.9	25.8	T_{NPC}	77.9	25.8	0
E	T_{in}	83.5	T_{out}	69.7	16.7	T_{in}	70.5	16.1	-0.6

Table A.4 Simulation parameters for a 3.3 kV 3L-ANPC VSC at 750 Hz

	Operating points			Simulation parameters of a 3.3 kV 3L-NPC VSC 4.5 kV PP Westcode T1200EB45E IGBTs and 4.5 kV PP Infineon D1031SH45T diodes					
	pf	m_a	PWM scheme	f_{sw} in Hz	$i_{ph,rms}$ in A	f_1 in Hz	U_{dc} in V	ϑ_a in °C	$\vartheta_{j,max}$ in °C
A	+1	1.15	3L-SVM		1040				
B	−1	1.15	3L-SVM	750	1040	50	5000	50	125
C	±1	0.05	2L-SVM		1200				

Table A.5 Percentage decrease of the maximal junction temperature

	NPC		ALB			PALB			
	Power devices	ϑ_{max} in °C	Power devices	ϑ_{max} in °C	p_{max} ALB in %	Power devices	ϑ_{max} in °C	p_{max} PALB in %	Gain in %
A	T_{out}	124.2	T_{in}	102.9	17.2	T_{in}	102.4	17.6	0.4
B	T_{in}	99.1	D_{in}	84.0	15.2	D_{out}	84.3	14.9	-0.3
C	T_{in}	109.7	T_{out}	87.8	19.9	T_{in}	86.5	21.1	1.2

131

Appendix

Table A.6 Percentage decrease of the average junction temperature

		NPC		ALB			PALB			Gain in %
		Power devices	$\vartheta_{j,avg}$ in °C	Power devices	$\vartheta_{j,avg}$ in °C	p_{max} ALB in %	Power devices	$\vartheta_{j,avg}$ in °C	p_{max} PALB in %	
A		T_{out}	112.1	T_{in}	93.5	16.5	T_{in}	93.1	16.9	0.4
B		T_{in}	90.9	D_{in}	79.6	12.4	D_{out}	79.7	12.3	-0.1
C		T_{in}	96.7	T_{out}	79.3	17.9	T_{in}	78.3	18.9	1

Table A.7 Simulation parameters for a 3.3 kV 3L-ANPC VSC at 1050 Hz

		Operating points		Simulation parameters of a 3.3 kV 3L-NPC VSC 4.5 kV PP Westcode T1200EB45E IGBTs and 4.5 kV PP Infineon D1031SH45T diodes						
	pf	m_a	PWM scheme	f_{sw} in Hz	$i_{ph,rms}$ in A	f_1 in Hz	U_{dc} in V	ϑ_a in °C	$\vartheta_{j,max}$ in °C
A	+1	1.15	3L-SVM		850				
B	−1	1.15	3L-SVM	1050	850	50	5000	50	125
C	±1	0.05	2L-SVM		1200				

Table A.8 Percentage decrease of the maximal junction temperature

		NPC		ALB			PALB			Gain in %
		Power devices	ϑ_{max} in °C	Power devices	ϑ_{max} in °C	p_{max} ALB in %	Power devices	ϑ_{max} in °C	p_{max} PALB in %	
A		T_{out}	124.8	T_{in}	99.2	20.5	T_{in}	98.9	20.7	0.2
B		T_{in}	105.8	D_{in}	80.5	23.9	D_{in}	80.4	24.0	0.1
C		T_{in}	121.9	T_{in}	96.3	21.1	T_{in}	93.9	23.0	1.9

Table A.9 Percentage decrease of the average junction temperature

		NPC		ALB			PALB			Gain in %
		Power devices	$\vartheta_{j,avg}$ in °C	Power devices	$\vartheta_{j,avg}$ in °C	p_{max} ALB in %	Power devices	$\vartheta_{j,avg}$ in °C	p_{max} PALB in %	
A		T_{out}	113.0	T_{in}	90.7	19.7	T_{in}	90.5	19.9	0.2
B		T_{in}	97.1	D_{in}	76.9	20.8	D_{in}	76.7	21.0	0.2
C		T_{in}	105.5	T_{in}	85.9	18.6	T_{in}	84.9	19.4	0.8

B Thermal behavior of a 3.3 kV 3L-ANPC VSC using 4.5 kV IGBT modules

The performances of a 3.3 kV 3L-ANPC VSC using 4.5 kV IGBT modules is further evaluated at grid frequency applying the PALB method and the ALB method, respectively. Tables B1, B.4, and B.7 summarize the simulations parameters.

Tables B.2, B.3, B.5, B.6, B.8, and B.9 summarize the decrease of the maximal and average junction temperature applying the balancing methods in comparison to the conventional NPC converter. The PALB method present substantial reduction of the junction temperature in comparison to the 3L-NPC VSC. However, the ALB method generally presents a better performance than the PALB method.

Table B.1 Simulation parameters for a 3.3 kV 3L-ANPC VSC at 500 Hz (SHE-5α)

Operating points			Simulation parameters of a 3.3 kV 3L-ANPC VSC 4.5 kV Mitsubishi CM900HB90H IGBT module					
pf	m_a	PWM scheme	f_{sw} in Hz	$i_{ph,rms}$ in A	f_1 in Hz	U_{dc} in V	ϑ_a in °C	$\vartheta_{j,max}$ in °C
A	+1	1.15	SHE-5α		480				
B	−1	1.15	SHE-5α	500	480	50	4854	37	67
C	±1	0.05	2L-SVM		610				

Table B.2 Percentage decrease of the maximal junction temperature

	NPC		ALB			PALB			
	Power devices	ϑ_{max} in °C	Power devices	ϑ_{max} in °C	p_{max} ALB in %	Power devices	ϑ_{max} in °C	p_{max} PALB in %	Gain in %
A	T_{out}	68.8	T_{in}	60.1	12.6	D_{in}	60.5	12.5	-0.1
B	D_{in}	66.9	D_{in}	60.5	9.6	D_{in}	64.6	3.4	-6.2
C	T_{in}	68.3	D_{in}	61.4	10.6	D_{in}	62.1	9.4	-1.2

Table B.3 Percentage decrease of the average junction temperature

	NPC		ALB			PALB			
	Power devices	$\vartheta_{j,avg}$ in °C	Power devices	$\vartheta_{j,avg}$ in °C	p_{max} ALB in %	Power devices	$\vartheta_{j,avg}$ in °C	p_{max} PALB in %	Gain in %
A	T_{out}	67.4	T_{in}	59.0	12.3	D_{in}	59.2	12.3	0
B	D_{in}	65.7	D_{in}	59.2	9.8	D_{in}	63.4	3.5	-6.3
C	T_{in}	66.4	D_{in}	58.7	11.7	D_{in}	60.2	9.4	-2.3

133

Table B.4 Simulation parameters for a 3.3 kV 3L-ANPC VSC at 750 Hz

				Simulation parameters of a 3.3 kV 3L-ANPC VSC 4.5 kV Mitsubishi CM900HB90H IGBT module					
		Operating points							
	pf	m_a	PWM scheme	f_{sw} in Hz	$i_{ph,rms}$ in A	f_l in Hz	U_{dc} in V	ϑ_a in °C	$\vartheta_{j,max}$ in °C
A	+1	1.15	SHE-5α		310				
B	−1	1.15	SHE-5α	750	310	50	4854	37	67
C	±1	0.05	2L-SVM		500				

Table B.5 Percentage decrease of the maximal junction temperature

	NPC		ALB			PALB			
	Power devices	ϑ_{max} in °C	Power devices	ϑ_{max} in °C	p_{max} ALB in %	Power devices	ϑ_{max} in °C	p_{max} PALB in %	Gain in %
A	T_{out}	68.6	T_{in}	57.3	16.5	T_{out}	57.3	16.4	-0.1
B	T_{in}	66.1	D_{in}	53.3	19.3	D_{in}	54.4	17.7	-1.6
C	T_{in}	69.5	T_{in}	59.3	14.5	D_{in}	60.9	12.2	-2.3

Table B.6 Percentage decrease of the average junction temperature

	NPC		ALB			PALB			
	Power devices	$\vartheta_{j,avg}$ in °C	Power devices	$\vartheta_{j,avg}$ in °C	p_{max} ALB in %	Power devices	$\vartheta_{j,avg}$ in °C	p_{max} PALB in %	Gain in %
A	T_{out}	67.2	T_{in}	56.2	16.3	T_{out}	56.3	16.2	-0.1
B	T_{in}	64.9	D_{in}	52.6	18.9	D_{in}	53.7	17.3	-1.6
C	T_{in}	66.9	T_{in}	57.8	13.6	D_{in}	59.7	10.8	-2.8

Table B.7 Simulation parameters for a 3.3 kV 3L-ANPC VSC at f_{sw}=1050 Hz (SVM)

				Simulation parameters of a 3.3 kV 3L-ANPC VSC 4.5 kV Mitsubishi CM900HB90H IGBT module					
		Operating points							
	pf	m_a	PWM scheme	f_{sw} in Hz	$i_{ph,rms}$ in A	f_l in Hz	U_{dc} in V	ϑ_a in °C	$\vartheta_{j,max}$ in °C
A	+1	1.15	SHE-5α		220				
B	−1	1.15	SHE-5α	1050	220	50	4854	37	67
C	±1	0.05	2L-SVM		400				

Table B.8 Percentage decrease of the maximal junction temperature

	NPC		ALB			PALB			
	Power devices	ϑ_{max} in °C	Power devices	ϑ_{max} in °C	p_{max} ALB in %	Power devices	ϑ_{max} in °C	p_{max} PALB in %	Gain in %
A	T_{out}	68.2	T_{in}	55.7	18.3	T_{in}	56.0	17.8	-0.5
B	T_{in}	66.5	T_{in}	51.9	21.9	T_{NPC}	52.0	21.8	-0.1
C	T_{in}	69.5	T_{out}	59.4	14.6	T_{out}	59.4	12.9	0

Table B.9 Percentage decrease of the average junction temperature

	NPC		ALB			PALB			
	Power devices	$\vartheta_{j,avg}$ in °C	Power devices	$\vartheta_{j,avg}$ in °C	p_{max} ALB in %	Power devices	$\vartheta_{j,avg}$ in °C	p_{max} PALB in %	Gain in %
A	T_{out}	66.8	T_{in}	54.8	17.9	T_{in}	55.1	17.5	-0.4
B	T_{in}	65.3	T_{in}	51.4	21.3	T_{NPC}	51.3	21.4	0.1
C	T_{in}	66.5	T_{out}	57.9	14.6	T_{out}	58.0	12.8	-0.1

Bibliography

[1] S. Bernet, "State of the Art and Developments of Medium Voltage Converters – An overview", Przeglad Elektrotechniczny (Electrical Review), vol. 82, no. 5, May 2006, pp. 1-10

[2] B. Wu, "High- Power Converters and AC Drives", Piscataway, New Jersey: Wiley Intersascience, IEEE press, 2006

[3] A. Nabae, I. Takahashi and H. Akagi, " A new neutral-point-clamped PWM inverter," IEEE Trans. Ind. Appl., vol. IA-17 , no.5, pp. 518-523, Sep./Oct. 1981

[4] S. Bernet, R. Sommer, "Multi-Level Converters for Industrial Applications", Vorlesungsskript TU Dresden, 2010

[5] J. Rodriguez, S. Bernet, P.K: Steimer, I.E. Lizama, "A Survey on Neutral-Point-Clamped Inverters," IEEE Transactions on Industrial Electronics, Vol. 57, Nr. 7, July 2010, pp. 2219-2230

[6] W. Srirattanawichaikul, Y. Kumsuwan , S. Premrudeepreechacharn, B. Wu, "A Vector Control of a Grid-Connected 3L-NPC-VSC with DFIG Drives," ECTI-CON, Int. Conf., Mai 2010

[7] M. Malinowski, S. Stynski, W. Kolomyjski, and M. P. Kazmierkowski, "Control of three-level PWM converter applied to variable-speed-type turbines," IEEE Trans. Ind. Electron., vol. 56, no. 1, pp. 69–77, Jan. 2009

[8] M. Saeedifard, H. Nikkhajoei, R. Iravani, "A Space Vector Modulated STATCOM Based on a Three-Level Neutral Point Clamped Converter," IEEE Trans. Power Delivery,vol 22, no. 2, April 2007, p. 1029-1039

[9] T. Brückner, "The active NPC Converter for Medium-Voltage Drives ," Achen Verlag Shaker, 2006

[10] T. Brückner and S. Bernet, "Loss balancing in the three-level voltage source inverters applaying active NPC switches ," in Proc. IEEE- PESC, Vancuver, Canada, 2001, pp. 1135-1140

[11] O. Apeldoorn, B. Ødegård, P. Steimer, S. Bernet, "A 16 MVA ANPC PEBB with 6 kA IGCTs ," Proc. IAS, Shanghai 2005

136

[12] N. Mohan, T.M. Undeland, W.P. Robbins, "Power electronics", John Wiley & Sons. INC. pp. 225-246

[13] S. Bernet, "Recent Developments of High Power Converters for Industry and Traction Applications", IEEE Transactions on Power Electronics, 2000, vol. 15, no.6, pp.1102-1117

[14] D. Andler, R. Alvarez, S. Bernet, J. Rodriguez, R. Sommer, "Experimental investigation of the commutation of the 3L-ANPC VSC phase leg using a 5.5 kA-IGCTs" in Industrial Electronics, IEEE Transactions on, Nov. 2013, pp 4820 - 4830

[15] C. Fink, " Untersuchung neuartiger Konzepte zur geregelten Ansteuerung von IGBTs", Südwestdeutscher Verlag für Hochschulschirften, 2010

[16] R. Alvarez, "Characterization of New Press-Pack IGBTs and Automated Delay Time Compensation of Parallel Connected IGBTs, 2011

[17] S. Bernet, E. Carroll, P. Streit, O. Apeldoorn, P. Steimer, and S. Tschirley, "Design, test and characteristics of 10-kV integrated gate commutated thyristors," IEEE Ind. Appl. Mag., vol. 11, no. 2, pp. 53–61, Mar./Apr. 2005

[18] R.D. Klug, N. Klaassen, "High Power Medium Voltage Drives-Innovations, Portfolio, Trends", Conf. Rec. of EPE, Dresden, Germany, 2005

[19] S. Rohner, "Untersuchung des Modularen Mehrpunktstromrichters M2C für Mittelspannungs-anwendungen", Dr. Hut Verlag, 2011

[20] W. Timpe, "Cycloconverters Drives for Rolling Mills," in IEEE Trans. Ind. Applicat. IA-18 (1982), July, Nr. 4, pp. 400-404

[21] M. Michel, "Leistungselektronik – Einfürung in Schaltungen und deren Verhalten," Berlin: Springer Verlag, 1971

[22] J. Kang, E. Yamamoto, M. Ikeda, E. Watanabe, "Medium Voltage Matrix Converter Design Using Cascaded Single-phase Power Cell Modules," IEEE Trans. on Industrial Electronics, Mai 2011

[23] YASKAWA: "Super Energy-Saving Medium Voltage Matrix Converter with Power Regeneration FSDrive-MX1S," Online www.yaskawa.de

Bibiography

[24] J. Rodriguez, S. Bernet, B. Wu, J. O. Pontt, S. Kouro "Multilevel Voltage-Source-Converter Topologies for Industrial Medium- Voltage Drives " IEEE Transaction on Industrial Electronics, vol. 54, no. 6, December 2007, pp. 2930-2945

[25] R. Emery, J. Eugene, "Harmonic losses in LCI-fed Synchronous motors", IEEE Trans. Ind. Appl., vol. 38, no.4, pp 948-954, Jul./Aug. 2002

[26] D. Krug, M. Malinowski, S. Bernet, "Design and Comparison of Medium Voltage Multi-Level Converters for Industry Application", in Conf. Rec. IEEE-IAS Annu. Meeting, 2004, vol.2, pp 761-790

[27] F. Kieferdorf, M. Basler, L.A. Serpa, J. H- Fabian, A. Coccia, G.A. Scheuer, " A new Medium Voltage Drive System Based on ANPC-5L Technology, " in Industrial Technology, IEEE International Conference on, March 2010, pp. 643-649

[28] S. Rohner, S. Bernet, M. Hiller, R. Sommer, "Modulation, Losses and Semiconductor Requirements of Modular Multilevel Converters," IEEE Trans. Ind. Electron., vol. 57, no. 8, pp. 2633–2642, Aug. 2010

[29] SIEMENS AG: "HVDC PLUS (VSC Technology)," www.energy.siemens.com

[30] R. H. Baker, "Bridge converter circuit," U.S. Patent 4 270 163, May 26, 1981

[31] X. del Toro Garcia, A. Arias, M. G. Jayne, and P. A. Witting, "Direct torque control of induction motors utilizing three-level voltage source inverters," IEEE Trans. Ind. Electron., vol. 55, no. 2, pp. 956–958, Feb. 2008

[32] A. Sapin, P. K. Steimer, and J.-J. Simond, "Modeling, simulation, and test of a three-level voltage-source inverter with output LC filter and direct torque control," IEEE Trans. Ind. Appl., vol. 43, no. 2, pp. 469–475, Mar./Apr. 2007

[33] M. Malinowski, M. Kazmierkowski, and A. Trzynadlowski, "A comparative study of control techniques for PWM rectifiers in AC adjustable speed drives," IEEE Trans. Power Electron., vol. 18, no. 6, pp. 1390–1396, Nov. 2003

[34] G. Abad, M. A. Rodriguez, and J. Poza, "Three-level NPC converterbased predictive direct power control of the doubly fed induction machine at low

138

constant switching frequency," IEEE Trans. Ind. Electron., vol. 55, no. 12, pp. 4417–4429, Dec. 2008

[35] R. Vargas, P. Cortes, U. Ammann, J. Rodriguez, and J. Pontt, "Predictive control of a three-phase neutral-point-clamped inverter," IEEE Trans. Ind. Electron., vol. 54, no. 5, pp. 2697–2705, Oct. 2007

[36] J. Holtz, "Pulsewidth modulation-A survey," IEEE Trans. Ind. Electron. , vol. 39, no. 5, pp. 410–420, Oct. 1992.

[37] J. Zaragoza, J. Pou, S. Ceballos, E. Robles, P. Ibaez, and J. L. Villate, "A comprehensive study of a hybrid modulation technique for the neutralpoint-clamped converter," IEEE Trans. Ind. Electron., vol. 56, no. 2, pp. 294–304, Feb. 2009

[38] A. Videt, P. Le Moigne, N. Idir, P. Baudesson, and X. Cimetiere, "A new carrier-based PWM providing common-mode-current reduction and dc-bus balancing for three-level inverters," IEEE Trans. Ind. Electron., vol. 54, no. 6, pp. 3001–3011, Dec. 2007

[39] S. Busquets-Monge, J. D. Ortega, J. Bordonau, J. A. Beristain, and J. Rocabert, "Closed-loop control of a three-phase neutral-pointclamped inverter using an optimized virtual-vector-based pulsewidth modulation," IEEE Trans. Ind. Electron., vol. 55, no. 5, pp. 2061–2071, May 2008

[40] J. I. Leon, S. Vazquez, R. Portillo, L. G. Franquelo, J. M. Carrasco, P. W. Wheeler, and A. J. Watson, "Three-dimensional feedforward space vector modulation applied to multilevel diode-clamped converters," IEEE Trans. Ind. Electron., vol. 56, no. 1, pp. 101–109, Jan. 2009.

[41] T. Brückner and D.G. Holmes, "Optimal pulse-width modulation for three-level inverters," IEEE Trans. Power Electron., vol. 20, no. 1, pp. 82–89, Jan. 2005.

[42] L. G. Franquelo, J. Napoles, R. C. P. Guisado, J. I. Leon, and M. A. Aguirre, "A flexible selective harmonic mitigation technique to meet grid codes in three-level PWM converters," IEEE Trans. Ind. Electron., vol. 54, no. 6, pp. 3022–3029, Dec. 2007.

Bibiography

[43] H. S. Patel and R. G. Hoft, "Generalized techniques of harmonic elimination and voltage control in thyristor inverters: Part I-Harmonic elimination," IEEE Trans. Ind. Appl., vol. IA-9, no. 3, pp. 310–317, May 1973

[44] J. Pontt, J. Rodriguez, and R. Huerta, "Mitigation of noneliminated harmonics of SHEPWM three-level multipulse three-phase active front end converters with low switching frequency for meeting standard IEEE- 519-92," IEEE Trans. Power Electron., vol. 19, no. 6, pp. 1594–1600, Nov. 2004

[45] S. Sirisukprasert, J.-S. Lai, and T.-H. Liu, "Optimum harmonic reduction with a wide range of modulation indexes for multilevel converters," IEEE Trans. Ind. Electron., vol. 49, no. 4, pp. 875–881, Aug. 2002

[46] J. Holtz and S. F. Salama, "Megawatt GTO-inverter with three-level PWM control and regenerative snubber circuit", in Proc. IEEE-PESC, Kyoto, JP,1988, pp1263-1270

[47] M. Bruckmann, R. Sommer, M. Fasching, and J. Sigg, "Series connection of high voltage IGBT modules," in Conf. Rec. IEEE-IAS Annu. Meeting, St. Louis, MO, 1998, pp 1067-1072

[48] R. Teichmann, T. Brückner and J. Mast, "Aspekte der Reichenschaltug von Leistungshalbleitern in der Mittelspannungstechnik," ELEKTRIE , Berlin 55 (2001) 05-07, pp. 328-331

[49] A. Nagel, S. Bernet, P. K. Steimer, and O. Apeldoorn, "A 24 MVA inverter using IGCT series connection for medium voltage applications," in Conf. Rec. IAS Annu. Meeting, Chicago, IL, 2001, pp. 867–870

[50] N. Celanovic and D. Boroyevich, "A comprehensive study of neutral-point voltage balancing problem in three-level-neutral-point-clamped voltage source PWM inverters," IEEE Trans. Power Electron., vol. 15, no. 2, pp. 242–249, Mar. 2000

[51] C. Newton and M. Sumner, "Neutral point control for multi-level inverters: Theory, design, and operational limitations," in Conf. Rec. IAS Annu. Meeting, New Orleans, LA, 1997, pp. 1136–1343

[52] Jung-Dae Lee, Tae-Jin Kim, Jae-Chul Lee, Dong-Seok Hyun, "A Novel Fault Detection of an Open-Switch Fault in the NPC Inverter System," Industrial Electronics Society, 2007 IEEE 33th Conference, pp. 1565-1569, Nov 2007

140

[53] Shengming Li, L. Xu, "Strategies of Fault Tolerant Operation for Three-Level PWM Inverters," IEEE Trans. on Power Electronics, vol 21, no. 4, pp.933-940, July 2006

[54] S. Ceballos, J. Pou, E. Robles, I.Gabiola, J. Zaragoza, J.L.Villate, "Three-Level Converter Topologies With Switch Breakdown Fault-Tolerance Capability," IEEE Trans. on Industrial Electronics, vol.55, no.3, pp.982-995, March 2008

[55] S. Bernet and T. Brückner, "Open-loop and closed-loop control method for a three-point converter with active switches, and apparatus for this purpose," US patent 6,697,274, Feb. 2004 (prority Sep. 2000)

[56] T. Brückner, S. Bernet, and H. Güldner, "The active NPC converter and its loss-balancing control," IEEE Trans. Ind. Electron., vol. 52, no. 3, pp. 855–868, Jun. 2005

[57] T. Brückner, S. Bernet, and P. K. Steimer, "Feedforward loss control of three-level active NPC converters," IEEE Trans. Ind. Appl., vol. 43, no. 6, pp. 1588–1596, Nov./Dec. 2007

[58] D. Floricau, E. Floricau, and G. Gateau, "Three-level active NPC converter: PWM strategies and loss distribution," in Proc. IEEE IECON, Nov. 2008, pp. 3333–3338

[59] D. Floricau, E. Floricau, and M. Dumitrescu, "Natural doubling of the apparent switching frequency using three-level ANPC converter," in Proc. ISNCC Conf. Rec., Lágow, Poland, Jun. 2008, pp. 1–6

[60] J. Sayago, S.Bernet, T. Brückner, "Comparison of medium Voltage IGBT-based 3L-ANPC-VSCs" in Proc. IEEE-PESC, Rhodes, Greece, 2008, pp 851-858

[61] D. Andler, E. Hauk, R. Alvarez, S. Bernet and J. Rodriguez, " New junction temperature balancing method for a Three Level Active NPC converter," EPE 2011

[62] A. Hämmerli, B. Ødegård, "AC excitation with ANPC," ABB Review 3/2008, www.abb.com

141

Bibiography

[63] J. Li, A. Q. Huang, S. Bhattacharya, and G. Tan, "Three-level active neutral-point-clamped (ANPC) converter with fault tolerant ability," in Proc. IEEE APEC, Feb. 2009, pp. 840–845

[64] S. R. Bowes and A. Midoun, "Suboptimal switching strategies for micrprocessor-controled PWM inverters drives,"IEE Proc. B, vol. 132, pp 133-148, May 1985

[65] S. Bernet, "Leistungshalbleiter als Nullstromschalter in Stromrichtern mit weichen Schaltvorgängen". Aachen: Verlag Shaker, 1995

[66] D.G. Holmes and T. Lipo, "Pulse Width Modulation for Power Converters," New York: Wiley, 2003

[67] F. Wang, "Sine-triangle vs. space vector modulation for Three-level PWM voltage source inverter," ieee Transaction on Industry Application, 2002, Vol. 38, No.2, pp. 500-506

[68] B.P. McGrath, D.G. Holmes and T. Lipo, "Optimized space vector switching sequence for multilevel inverters," in Proc. IEEE-APEC, Anahein, 2001, pp. 1123-1129, and IEEE Trans. Power Electron., vol.18, pp. 1293-1301, Nov. 2003

[69] Y.-H. Lee, D.-H. Kim and D.-S. Hyun, "Carrier based SVPWM method for multi-level system with reduced HDF," in Proc. IEEE-IAS Annu. Meeting, Rome, 2000, pp.1996-2003

[70] J. Sayago, "Investigation and comparison of three-level NPC converters for medium voltage application ," Dr. Hut Verlag, 2009

[71] X. Yuan, H. Stemmler and I. Barbi, "Investigation on the clamping voltage self-balancing of the three-level capacitor clamping inverter," in Proc. IEEE-PESC, Charleston, SC, 1999, pp. 1059-1064.

[72] R. Sommer, A. Mertens, M. Griggs, H.-J. Conraths, M. Bruckmann, T. Greif, "New Medium Voltage Drive Systems using Three-Level Neutral Point Clamped. Inverter with High Voltage IGBT," Industry Applications Conference, pp 1513-1519, IEEE 1999

[73] "Application Handbook," www.semikron.com, 2011 pp 127

142

[74] S. S. Fazel, "Investigation and Comparison of Multi-Level Converters for Medium Voltage Applications," 2007

[75] D.A. Murdock, J.E. Ramos, J.J. Connors, R.D. Lorenz "Active Thermal Control of Power Electronics Modules," in Conf. Rec. IEEE-IAS Annu. Meeting, Salt Lake City, UT, 2003, pp. 1511-1515

[76] J. Van den Keybus, T. Nobels, R. Belmans, "Thermal design of converters using discrete power components incorporating an IGBT and a freewheeling diode", in: Power Electronics and Applications, 2005

[77] T. Schütze, H. Berg, and O. Schilling, "The new 6.5 kV IGBT module: a reliable device for medium voltage application," PCIM 2001, Nürnberg, Germany

[78] A. Morozumi, K Yamada, T. Miyasaka, "Reliability design technology for power semiconductors module," www.fujielectric.com, Fuji Electric Review, Vol.47, 2001

[79] T. Schütze, "Hochleistungs-IGBTs in Traktion- und Industrie: Anwendungen und Anforderungen, Technologie und Zuverlässigkeit." In: Ansteuer- und Schutzschaltungen für Mosfet und IGBT ECPE-Cluster Leistungselektronik, 2010

[80] Lixiang Wie, R.J. Kerkman, R.A. Lukaszewski, B.P. Brown, N. Gollhardt, B.W. Weiss, "Junction Temperature Prediction of a Multiple-chip IGBT Module under DC Condition," in Industry Applications Conference, 2006, pp 754-762

[81] U. Drofenik, J.W. Kolar "Teaching thermal design of power electronics systems with Web-based interactive educational software," www.ipes.ethz.ch

[82] M. März, P. Nance "Thermal modeling of Power-electronic Systems", www.infineon.com

[83] F. Christiaens and E. Beyne, "Transient thermal modelling and characterisation of a hybrid component," in Electronic Components and Technology Conference, 1996, pp 154-164

[84] "Thermal equivalent circuit models ", AN2008-03, www.infineon.com

143

Bibiography

[85] Z. Jakopovid, Z. Bencic, R. Koncar, "Identification of thermal equivalent-circuit parameters for semiconductors," Computers in Power Electronics, 1990 IEEE Workshop

[86] V. Blasko, R. Lukaszewski, and R. Sladky, "On line thermal model and thermal management strategy of a three phase voltage source inverter," in Conf. Rec. IEEE-IAS Annu. Meeting, Phoenix, AZ, 1999, pp 1423-1431

[87]. Z. Luo; H. Ahn; M.A.E.Nokali: "A thermal model for insulated gate bipolar transistor module", in IEEE Trans. Power Electronics, Vol. 19, No. 4, pp 902-907, July 2004

[88] F.N. Masana "A straightforward analytical method for extraction of semiconductor device transient thermal parameters," ScienceDirect, November 2006

[89] T. Franke, G. Zaiser, J. Otto, M. Honsberg-Riedl, R. Sommer, "Current and temperature distribution in multi chip modules under inverter operation", in EPE Conf. Rec., Lausanne , Switzerland, 1999, CD-ROM

[90] S. Konrad, K. Anger, "Electro-thermal model for simulating chip temperatures in PWM inverters," in Power Conversion, Proceeding 219, June 1995

[91] D. Andler, "Experimental Investigation of Three-Level Active Neutral Point Clamped Voltage Source Converters using Integrated Gate-Commutated Thyristors", Verlag Dr. Hut, Mai 2014

[92] R. Alvarez, F. Filsecker, and S. Bernet, "Characterization of a new 4.5 kV press pack spt+ igbt for medium voltage converters," in Energy Conversion Congress and Exposition (ECCE), 2009 IEEE, (San Jose, USA), pp. 3954–3962, sep 2009

[93] R. Alvarez, S. Bernet, L. Lindenmüller, and F. Filsecker, "Characterization of a new 4.5 kV press pack spt+ igbt in voltage source converters with clamp circuit," in Industrial Technology (ICIT), 2010 IEEE International Conference on, (Valparaiso, Chile), pp. 702 – 709, mar 2010

[94] Datasheet CM 900HB 90H, Tech. Information IGBT-Modules, Powerex Corp, Pennsylvania, USA

[94] Datasheet press pack IGBT T1200EB45E IXYS UK Westcode, www.westcode.com

[95] Datasheet Diode D1031SH45 Infineon, www.infineon.com

[96] Y. Shakweh, "Critical assessment of HV power devices for MV PWM VSI converters", in EPE Conf. Rec., Lausanne , Switzerland, 1999, CD-ROM

www.ingramcontent.com/pod-product-compliance
Lightning Source LLC
Chambersburg PA
CBHW060314220326
41598CB00027B/4322